# METHODS in MICROBIOLOGY

# METHODS in

# MICROBIOLOGY

*Edited by*

**T. BERGAN**

*Department of Microbiology,*
*Institute of Pharmacy and Department of Microbiology,*
*Aker Hospital, University of Oslo,*
*Oslo, Norway*

**J. R. NORRIS**

*Agricultural Research Council,*
*Meat Research Institute,*
*Bristol, England*

# Volume 12

 1978

ACADEMIC PRESS
London · New York · San Francisco
*A Subsidiary of Harcourt Brace Jovanovich, Publishers*

ACADEMIC PRESS INC. (LONDON) LTD
24–28 Oval Road
London NW1

U.S. Edition published by
ACADEMIC PRESS INC.
111 Fifth Avenue
New York, New York 10003

Library of Congress Catalog Card Number: 68–57745
ISBN: 0–12–521512–6

PRINTED IN GREAT BRITAIN BY
ADLARD AND SON LIMITED
DORKING, SURREY

# LIST OF CONTRIBUTORS

T. BERGAN, *Department of Microbiology, Institute of Pharmacy and Department of Microbiology, Aker Hospital, University of Oslo, Oslo, Norway*

H. BRANDIS, *Institute of Medical Microbiology and Immunology, University of Bonn, West Germany*

J. HENRICHSEN, *The Streptococcus Department, Statens Seruminstitut, Copenhagen, Denmark*

S. HENRIKSEN, *Kaptein Wilhelmsen og Frues Bakteriologiske Institutt, University of Oslo, National Hospital of Norway*

J. JELINKOVA, *Postgraduate Medical and Pharmaceutical Institute, Department of Medical Microbiology, Prague*

ERNA LUND, *The Pneumococcus Department, Stratens Seruminstitut, Copenhagen, Denmark*

S. MUKERJEE, *WHO International Reference Centre for Vibrio Phage-Typing, Calcutta*

P. OEDING, *The Gade Institute, Department of Microbiology, University of Bergan, Bergan, Norway*

J. ROTTA, *Institute of Hygiene and Epidemiology, Prague*

E. THAL, *National Veterinary Institute, Stockholm, Sweden*

S. WINBLAD, *Department of Clinical Bacteriology, University of Lund, Malmo General Hospital, Malmo, Sweden*

# PREFACE

Volume 12 of "Methods in Microbiology" continues the series of four Volumes which describe the methods available for typing the major human pathogens.

Volume 12 begins with a general article on the serotyping of bacteria and then presents three short Chapters on the characterisation of *Yersinia enterocolitica*. Two Chapters deal with *Vibrio cholerae* and the rest of the Volume is concerned with the Gram-positive cocci including Chapters on *Staphylococcus*, haemolytic streptococci, enterococci, bacteriocin typing of streptococci and the identification and characterisation of *Streptococcus pneumoniae*.

As with Volumes 10 and 11, we consider ourselves fortunate in having recruited internationally recognized authorities on these various topics as authors. We believe that the resulting compendium of methods, together with the detailed discussion of the epidemiological backgrounds, will provide a valuable reference work for microbiologists working in the epidemiological field and a valuable orientation for newcomers to this important area of microbiology.

T. Bergan
J. R. Norris

*October*, 1978

# CONTENTS

# CONTENTS OF PUBLISHED VOLUMES

# CHAPTER I

# Serotyping of Bacteria

## S. D. Henriksen

*Kaptein W. Wilhelmsen og Frues Bakteriologiske Institutt,*
*University of Oslo, National Hospital of Norway*

## I. INTRODUCTION

The use of serological methods for epidemiological typing is based upon the fact that micro-organisms frequently show variations in antigenic constitution, not only between distantly related or unrelated organisms, but even within groups of closely related organisms. The identification of the antigens of micro-organisms by means of specific antibodies, therefore, often allows very fine subdivision of microbial taxa, and greatly facilitates the surveillance and control of the spread of pathogenic micro-organisms.

Micro-organisms produce a wide variety of antigens, many of which can be utilised in identification and epidemiological typing: structural components of the cells, such as cell wall constituents, capsules or envelopes, flagellae, fimbriae; secretion products of the cells such as exotoxins or extracellular enzymes; or antigens contained in the interior of the cells. Many of these antigens can be utilised for typing purposes.

Chemically the antigens used for such purposes are of two main kinds: proteins and carbohydrates (including compounds composed of both polypeptide and carbohydrate components). Antigens are generally large molecules of complex structure, and, therefore, they frequently have more than one substructure which can serve as an antigenic determinant. Immune sera against antigens accordingly may contain separate antibodies against the various antigenic determinants on the same molecule, and even antibodies reacting with the same determinant may be heterogeneous molecules with different specificities, i.e. different degrees of complementarity to the same determinant.

## II. SPECIFICITY

The basis of the use of serological reactions is their specificity, the fact that antibodies produced in response to immunisation with one particular antigen give maximal reaction with that antigen only, i.e. a substance, or a substructure (a determinant) on an antigen with identical chemical constitution.

## III. CROSS-REACTIONS AND THEIR SIGNIFICANCE

In the early stages of the development of immunology the specificity of serological reactions was thought to be nearly absolute, with only few known exceptions. But accumulated experience has shown that this is not so. Serological cross-reactions are known to be of common occurrence and must be taken into consideration in the evaluation of the results of serological tests. With respect to cross-reactions there is a principal difference between protein antigens (i.e. antigens where the determinants are of protein nature) and carbohydrate antigens (antigens where the determinants are of carbohydrate nature).

In the case of carbohydrate antigens, cross-reactions are quite common and often occur between antigens from entirely unrelated organisms. Such cross-reactions are due to structural similarities, which may be quite accidental, such as shared monosaccharides in similar linkages. The principles of such cross-reactions have been eminently elucidated by Heidelberger and his co-workers in a large number of studies, and by other authors. As a consequence of such studies it is clear that cross-reactions between polysaccharide antigens only indicate structural similarity and do not necessarily indicate genetic relationship of the organisms producing the antigens.

With protein antigens the situation is quite different. As a general rule protein antigens do not cross-react with other proteins. But there is one very important exception: homologous, iso-functional proteins (i.e. of common phylogenetic origin and with identical function, e.g. enzymes) from different organisms may show cross-reactions (Cocks and Wilson, 1972; Gordon et al., 1969; Holten, 1974; Murphy and Mills, 1969; Steffen et al., 1973) which may vary in strength according to the degree of phylogenetic relationship between the organisms and to the rate of evolution of the protein in question, i.e. the frequency of mutational changes of the amino-acid sequence. Such cross-reactions can be utilised in measuring phylogenetic relationships between organisms. Cross-reactions between proteins due to accidental structural similarity appear to be unknown.

## IV. GENETIC DETERMINATION OF ANTIGENS

Many antigens are genetically determined by structural genes in the genome of the organism and are reasonably constant. But some antigens are determined by other, extraneous genetic elements, e.g. bacteriophages or by plasmids or episomes. Thus protein antigens such as exotoxins (Barksdale et al., 1961; Zabriskie, 1964) and O-antigen factors in Gram-negative bacteria (Iseki and Kashiwagi, 1955; Iseki and Sakai, 1954; Le Minor, 1963a, b, 1965a, b, c, 1966a, b, 1968; Le Minor et al., 1963; Le Minor et al., 1961; Staub and Forest, 1963) may appear or disappear with the entry or exit of a prophage from the genome. Antigens may also be modified by mutation, e.g. the S–R dissociation in enteric Gram-negative rods, which alters the O-antigen by eliminating side chains from its "core" polysaccharide (Kauffmann, 1961).

## V. DEMONSTRATION OF ANTIGENS

Antigens with only a single known determinant, eliciting antibodies of only one specificity, e.g. Salmonella O-antigens 11 or 16, can be demonstrated by the use of an untreated immune serum from a rabbit. But when the antigen is a mosaic of several determinants, each eliciting production of a separate antibody, it may be necessary to prepare factor sera, reacting only with one determinant, by absorbing the serum with organisms which possess one or more, but not all, of the same determinants. The antibodies which react with the determinants on the absorbing antigen are thus removed, whereas the antibodies which fail to react with any determinant on the absorbing antigen, remain in the serum. For example antiserum against Salmonella paratyphi A O-antigen (O: 1, 2, 12) contains antibodies anti-1, anti-2 and anti-12. By absorption with the *durazzo* variety of *S. paratyphi A* (O: 2, 12) the two latter antibodies are removed and only anti-1 remains. Many other examples may be found in Kauffmann (1961).

## VI. INTERPRETATION OF SEROLOGICAL REACTIONS

When an unknown microbial antigen reacts with an immune serum of known specificity, this may in principle be explained in two different ways: (a) either the unknown antigen may be identical with the one used to raise the immune serum, in which case titres of the serological reactions usually will be equal with the unknown and the homologous antigens, or (b) it may be a cross-reaction, in which case the titres given by the unknown antigen may be the same as, or lower than with the homologous antigen. In order to choose between the two alternatives, it may be necessary to carry out absorption of the serum with the unknown antigen. If this leads to complete exhaustion of the serum of all antibodies reacting with the homologous and the unknown antigens, the two antigens can be considered identical. But if only the antibody reacting with the unknown antigen is removed, whereas some of the antibody reacting with the homologous antigen remains, the two antigens are not identical.

In the evaluation of serological reactions it is important to realize that cross-reactions can be of two kinds: They may be due to sharing by the two antigens of one or more, but not all, identical antigenic determinants (e.g. O-antigens 1, 4, 5, 12 and 1, 9, 12 in the *Salmonella* group), or to similar, but not identical chemical structure of one or more determinants on the two antigens (e.g. the capsular polysaccharides of types 3 and 8 of *Streptococcus pneumoniae*, the former being a polymer of cellobiuronic acid, and the latter containing cellobiuronic acid units separated by one glucose and one galactose residue). Cross-reactions can be of all degrees, and they can be bilateral or unilateral.

## VII. KINDS OF SEROLOGICAL REACTIONS

A large variety of serological reactions are available. The reactions most commonly used for epidemiological typing are briefly presented in the following.

### A. Simple reactions requiring only antigen and antibody

#### 1. *The precipitation reaction*

This is a reaction between antigen in a colloidal solution and antibody, where lattices are formed and give visible precipitation. The lattice formation is due to the fact that each antibody molecule has two or more antigen binding sites, "valencies", and the antigen molecules, likewise, have several determinants to which antibody molecules can be bound and therefore the antigen functions as a multivalent substance. Variations in the relative proportions of antigen to antibody cause continuous variation in the composition of the precipitate. In the region of antibody excess the precipitate tends to be white, compact, dehydrated and insoluble. Towards the zone of antigen excess the precipitate becomes more translucent and gelatinous, more highly hydrated and increasingly soluble, until, in extreme antigen excess only soluble antigen-antibody complexes are formed. The composition of the precipitate is to some extent dependent upon the ionic strength of the solution.

The reaction exists in several modifications:

(a) *The simple ring test.* The antigen solution (often in ten-fold dilutions) is layered over immune serum in narrow tubes or capillaries. In positive reactions a disc of precipitate forms at the interface. The reaction is useful for qualitative demonstration or semi-quantitative measurement of antigens. Grouping and typing of streptococci is an example.

(b) *The quantitative precipitin reaction,* where precipitates formed by known volumes of antigen and serum are washed and the nitrogen content measured chemically. The reaction is not much used for typing purposes, but finds wide application as a research tool.

(c) *The capsular swelling reaction.* Dilute suspensions of encapsulated organisms, mostly bacteria, are mixed with serial dilutions of specific immune serum on slides. In positive cases antibody is bound to the outer layers of the capsules, which then become visible in the microscope, giving the impression that the capsule has swelled. The titre, the highest serum dilution giving a positive reaction, may be useful in distinguishing between a homologous reaction and a cross-reaction. The reaction is used for example, in typing of *Klebsiella* and *Haemophilus.*

(d) *Gel diffusion methods.* In single diffusion an antigen solution is layered over an agar column containing immune serum in a tube or capillary. When antigens and corresponding antibodies after diffusion meet in optimal proportions in the gel, precipitate bands appear, one or more according to the number of antigen–antibody systems involved. This reaction also finds little general application in typing. Double diffusion in tubes may similarly be set up: a layer of immune serum in the bottom, then a layer of clear agar and on top a layer of antigen solution. After diffusion precipitate bands form in the agar layer.

In double diffusion in agar layers, holes ("wells") are made in an agar layer in a Petri dish or on a glass plate. The number and arrangement of the wells can be varied at will, but the distances between the wells must always be such that visible precipitates can be formed between antigen wells and antiserum wells. Some wells are filled with antigen solutions, others with serum, or antibody solutions, and the dishes are incubated in a humid atmosphere until precipitates have appeared. Again one or more precipitate bands may form between an antigen well and a serum well according to the number of antigen–antibody systems. The method permits comparison of antigens against the same antibody (or vice versa), and it is possible to distinguish reactions of identity, cross-reactions or reactions of non-identity. A technique using cellulose acetate membranes instead of agar layers has also been developed. This method can be applied to many different problems, for example typing of capsular antigens or of viruses.

(e) More sophisticated methods are also available, and could be utilised for typing purposes: immunoelectrophoresis in agar cells or on cellulose acetate membranes, radio-immunoelectrophoresis, two-dimensional electrophoresis. A description of such methods would be outside the scope of this Chapter.

## 2. *The agglutination reaction*

This is a reaction between an antigen in the form of a suspension of particles and antibody. In positive reactions, the particles are bound together by antibody molecules with two or more antigen binding sites. The reaction may be carried out as a direct agglutination, where the antigen is a suspension of the organism in question: bacteria, *Rickettsia*, fungal elements. Such reactions may be carried out as qualitative tests, usually on slides, by suspending organisms from a culture on solid medium in a drop of serum, or serum dilution. Tests for demonstration of bacterial antigens such as *Salmonella* H- or O-antigens are usually carried out in this manner. The sera must have been carefully checked beforehand for reactivity and specificity, and for determination for the most suitable test dilution.

Agglutination may also be carried out in test-tubes by mixing serial (mostly two-fold) dilutions of serum with equal volumes of antigen. According to the nature of the antigen, incubation time and temperature may be varied. Agglutination of bacterial H-antigens by the corresponding antibody proceeds comparatively rapidly and is usually completed in 2–4 h. O-agglutination is slower and requires incubation overnight to be completed. The tests may be incubated in a water bath at 37 or 50 °C.

Agglutination reactions may also be carried out as indirect or "passive" agglutination, where the antigen, usually extracted from an organism, is bound to an inactive particle, for example latex or bentonite, or to erythrocytes (indirect haemagglutination). Antigen may be bound to untreated erythrocytes (lipopolysaccharides), to tanned corpuscles (proteins) or be fixed to the corpuscles by means of chromium chloride (certain polysaccharides). Before carrying out the test it may be necessary to absorb the serum with untreated erythrocytes to remove possible natural antibodies.

Indirect agglutination reactions are usually more sensitive than precipitation reactions with the same antigens. The indirect haemagglutination reaction in particular has been applied to numerous problems, including demonstration of bacterial antigens.

The titrations can be carried out by the conventional technique in tubes, or on plastic trays by the microtitre technique introduced by Takatsy (1955).

## B. Reactions requiring one or more additional components, e.g. complement or indicator systems

### 1. *Cytolysis–bacteriolysis*

When certain cells such as erythrocytes or bacteria e.g. *Vibrio cholerae* or some other Gram-negative rods, are exposed to specific antibodies and complement, the cell membranes are damaged and the cells lysed (the lysis of *Leptospira* cells by antibody is complement independent and does not belong in this category). If erythrocytes, coated with antigens on their surface are exposed to antibodies against the coating antigen and complement, lysis may also occur. Such reactions might be applied to typing, but apparently are not much used for such purposes at the present time.

### 2. *Complement fixation*

The formation of antigen-antibody complexes in the presence of complement causes activation of complement. One of the consequences of complement activation is the disappearance of its haemolytic activity. This is called "complement fixation". The loss of haemolytic activity may, therefore, be taken as evidence that an antigen–antibody reaction has occurred.

The reaction is carried out in two steps. In the first step a mixture of an antigen (known or unknown), a serum (unknown or known respectively) and complement is incubated in order that the antigen-antibody reaction and subsequent complement activation may take place. In the second step a test for remaining haemolytic complement activity is carried out by adding "sensitised" sheep erythrocytes (erythrocytes "coated" by rabbit antibody against sheep erythrocytes, immune haemolysin or "amboceptor"). If haemolysis occurs, it means that haemolytic complement still is available, complement has not been activated, the reaction is negative. Absence of haemolysis indicates that activation of complement has taken place, which, provided that all controls are satisfactory, indicates an antigen–antibody reaction, i.e. the reaction is positive. The reaction can be carried out by the conventional test-tube titration, or as a micro-modification in plastic trays with microtitre equipment according to Takatsy (1955).

Many different modifications of the test have been used, and it would be outside the scope of this Chapter to describe them. Only a few comments on the principles of the test are given here.

(a) *Complement.* The usual source of complement is fresh guinea pig serum. Guinea pig blood is drawn by heart puncture. As soon as possible after coagulation and clot retraction, the serum is separated from the clot and centrifuged. If it is to be used within 24 h, it may be stored at *c.* 4 °C. If not, it may be maintained in an active state by storage below −20 °C or by lyophilisation. When complement is being used to set up tests, it should preferably be kept cold in ice water to avoid inactivation at high room temperature.

In crude routine tests 0·85% saline is often used as diluent, but due to the fact that two of the steps in the complement activation require the presence of calcium ions and magnesium ions, respectively, saline does not ensure optimal complement activity, as the concentrations of these ions may be too low. It is preferable, therefore, to use a barbiturate buffer, as described by Mayer (1961), or a tris buffer as described by Levine (1967), which contain optimal amounts of these ions.

Because complement is a very unstable substance, a titration of its activity should be carried out on each occasion when it is going to be used. This is done by testing varying quantities of complement against constant quantities of immune haemolysin and of sheep erythrocytes. If the degree of haemolysis is plotted against the quantity of complement, a sigmoid curve is obtained, with the steepest slope corresponding to about 50% haemolysis. The most exact way to measure complement activity, therefore, is to determine the dilution that produces exactly 50% haemolysis. Deter-mination of the dilution that produces 100% haemolysis (which is often done), is much less accurate. In a complement fixation test it is necessary to use a slight excess of complement, as some of the other reagents used may have a slight anticomplementary effect. It is customary, therefore, to use an arbitrary number of 50% haemolytic units per unit volume, e.g. 4, 50% units. In complement fixation tests one must always include a complement control. i.e. a control with complement alone, without serum or antigen, to make sure that complement activity has been maintained during the test period.

(b) *Serum.* In complement fixation tests for typing purposes the serum will generally be an immune serum of known specificity. In order to make sure that the serum does not contain any complement, it is always in-activated, heated in a water bath at 56 °C for 30 min, before use. Serial dilutions, often two-fold, are usually tested against constant quantities of antigen and complement. A serum control, a tube with serum and complement, but without antigen, should always be set up to make sure that the serum does not contain anticomplementary substances.

(c) *Immune haemolysin.* This is usually raised in rabbits by giving them several injections of suspensions of washed sheep erythrocytes intra-venously, e.g. twice a week for 3 or 4 weeks. IgM antibodies, which appear early in the course of immunisation, are more active in immune haemolysis than the later appearing IgG antibodies, and it is therefore preferable not to continue the immunisation longer than necessary to get a good haemolytic titre. The serum is inactivated before use to get rid of any complement it may contain. The serum may be preserved with merthiolate 0·01% or sodium azide 0·1%, and will keep for a very long time at 4 °C, or it may be stored frozen below $-20$ °C. Each new serum has to be titrated to determine the titre and dilution to be used in complement fixation tests. The titration is carried out by testing twofold dilutions (usually starting at 1:100) of the serum against constant quantities of sheep erythrocytes and complement. As with complement, a moderate excess of haemolysin is also needed in the complement fixation reaction, e.g. two to four times the concentration needed to produce complete haemolysis in the titration.

(d) *Antigen.* A good antigen should have a high capacity for reacting with specific antibody and it should not be anticomplementary by itself. For this reason antigen to be used in a complement fixation test should be titrated against constant quantities of complement, but without antibodies, in order to determine whether it is anticomplementary or not. If it has an anti-complementary effect in higher concentration, it is important to select a dilution well below the anticomplementary level, e.g. not more than 1/4 of the lowest anticomplementary concentration. In the actual test a control with antigen and complement, but without serum should always be included.

(e) *Sheep erythrocytes.* Sheep blood is drawn into a sterile modified Alsever's solution (Mayer, 1961) and stored in the refrigerator at *c.* 4 °C. It is recommended that the blood should be left in the refrigerator for about

1 week before use, for stabilisation of its susceptibility to haemolysis. After the lapse of this period of time, the erythrocytes can be used for about 2 months.

Before use a suitable quantity of blood is centrifuged and plasma and buffy coat are pipetted off. The cells are then washed three times in 5–10 volumes of the isotonic diluent (Mayer, 1961). After the final centrifugation a suspension may be prepared by suspending 2–4% packed cells in saline (v/v), or, in more critical work, the cell suspension may be standardised spectrophotometrically to a selected optical density.

Complement fixation tests can be carried out by testing serial (usually twofold) serum dilutions against constant quantities of antigen and complement, or it may be carried out as checker board titrations by testing serial dilutions of serum against serial dilutions of antigen, and with constant complement.

### 5. *Neutralisation tests, requiring an indicator system*

The indicator system may be experimental animals as in exotoxin neutralisation tests (e.g. typing of toxins of *Clostridium botulinum*), cell cultures (e.g. virus neutralisation), leucocytes (opsonin tests), guinea pigs (passive cutaneous anaphylaxis). Not all these tests are at present being used for epidemiological typing, but they are potentially useful.

### C. Reactions using labelled reagents

By the use of fluorescent antibodies (labelled with substances like Fluorescein or Rhodamine-lissamine) it is possible to demonstrate microbial antigens. Antibodies labelled with ferritin or with peroxidase can be utilised for demonstrating microbial antigens by electron microscopy.

## VIII. CONTROLS

In order to get reliable and reproducible results of serological tests, it is essential that controls are included in the tests: controls with known antigens in tests designed to demonstrate antigens in unknown specimens, controls of the activity of immune sera, controls of all reagents which are subject to variation: erythrocytes, complement, etc.

## IX. PREPARATION AND HANDLING OF IMMUNE SERA

So many different methods have been used for the production of immune sera, that it is impossible to suggest a single method. Many methods and schedules of immunisation have been successful.

In general the rabbit is the experimental animal of choice, unless special reasons speak in favour of using a different animal.

Some antigens, e.g. bacterial suspensions, are very effective and simple intravenous injections in the ear veins give good results. Schedules such as two or three injections per week for 3 or 4 weeks are usually successful. Test bleedings from the ear may be taken 4 or 5 days after the last injection, and the animal may be bled from the heart the following day if the test bleeding proves satisfactory. If not, additional injections may be given until satisfactory titres of antibodies are obtained. It is important to select antigen doses that are non-toxic to the animals, otherwise it is impossible to recommend any particular dosage.

With some antigens, e.g. some protein antigens, it is less easy to produce potent immune sera. In these cases adjuvants may be used. The antigen may, for example, be emulsified in Freund's adjuvant (commercially available). Various schedules have been successful. In some cases a single dose, divided between four or five intramuscular sites is sufficient for good antibody production. In other cases the intramuscular injection of antigen in complete Freund's adjuvant may be followed about 4 weeks later by a new injection of antigen in incomplete Freund's adjuvant, or by an intravenous injection. Additional intravenous injections may be given at intervals of a few weeks.

Blood from immunised rabbits is drawn by heart puncture. The serum, after separation from the clot and centrifugation may be preserved with merthiolate (0·01%) or sodium azide (0·1%). Immune sera may be stored in the refrigerator at c. 4°C, and will keep for a very long time (often for a large number of years) if bacterial or fungal growth is avoided. Sera should always be handled using aseptic techniques. Sera may also be preserved by freezing below −20°C or by lyophilisation.

## X. IMPORTANCE OF THE NATURE OF THE ANTIBODIES

The nature of the antibodies is of some importance in serological tests. Antibodies of the IgM class are much more active than IgG, give higher titres, in certain reactions such as the agglutination reactions or in immune haemolysis. The specificity of the antibodies also appears to vary with the antibody class. Thus experience has indicated that antibodies of the IgM class, e.g. in horse sera against pneumococcus polysaccharides, have a broader specificity, giving more cross-reactions than antibodies of IgG, e.g. rabbit antisera against the same antigens.

## XI. SENSITIVITY OF SEROLOGICAL REACTIONS

An important consideration in the selection of serological reactions for specific purposes is the sensitivity of the reactions, i.e. how much antibody is required to give a positive reaction. In certain cases it is very important to select a highly sensitive method, in other cases this point is not of critical importance. The following list gives an approximative comparison of the sensitivities of some reactions (Humphrey and White, 1963):

| Reaction | Sensitivity in $\mu$g antibody N/ml required for positive reaction |
|---|---|
| Precipitation | 10 |
| Bacterial agglutination | 0·01 |
| Indirect haemagglutination | 0·003 |
| Haemolysis | 0·0002–0·03 |
| Complement fixation | 0·1 |
| Toxin neutralisation | 0·01 |
| Passive cutaneous anaphylaxis | 0·003 |

Details of these and other tests, as they are applied to specific diagnostic problems are to be found in the various Chapters in this Series, which deal with the typing of particular bacterial species.

## REFERENCES

Barksdale, L., Garmise, L., and Rivera, R. (1961). *J. Bact.*, **81,** 527–540.
Cocks, G. T., and Wilson, A. C. (1972). *J. Bact.*, **110,** 793–802.
Gordon, J., Schweiger, M. S., Krisko, I., and Williams, C. A. (1969), *J. Bact.*, **100,** 1–4.
Holten, E. (1974). *Acta path. microbiol. scand. Sect.* B, **82,** 849–859.
Humphrey, J. H., and White, R. G. (1963). "Immunology for Students of Medicine", p.192. Blackwell Scientific Publications, Oxford.
Iseki, S., and Kashiwagi, K. (1955). *Proc. Jap. Acad.*, **31,** 558–563.
Iseki, S., and Sakai, T. (1954). *Proc. Jap. Acad.*, **30,** 1001–1003.
Kauffmann, F. (1961) "Die Bakteriologie der *Salmonella* Species". Munksgaard, Copenhagen.
Le Minor, L. (1963a). *Ann. Inst. Pasteur*, **103,** 684–706.
Le Minor, L. (1963b). *Ann. Inst. Pasteur*, **105,** 879–896.
Le Minor, L. (1965a). *Ann. Inst. Pasteur*, **108,** 805–811.
Le Minor, L. (1965b). *Ann. Inst. Pasteur*, **109,** 35–46.
Le Minor, L. (1965c). *Ann. Inst. Pasteur*, **109,** 505–515.
Le Minor, L. (1966a). *Ann. Inst. Pasteur*, **110,** 380–386.
Le Minor, L. (1966b). *Ann. Inst. Pasteur*, **110,** 562–567.
Le Minor, L. (1968). *Ann. Inst. Pasteur*, **114,** 48–62.
Le Minor, L., Ackermann, H. W., and Nicolle, P. (1963). *Ann. Inst. Pasteur*, **104,** 469–476.
Le Minor, L., Le Minor, S., and Nicolle, P. (1961). *Ann. Inst. Pasteur*, **101,** 571–589.

Levine, L. (1967). In "Handbook of Experimental Immunology" (D. M. Weir, Ed.), p. 708. Blackwell Scientific Publications, Oxford and Edinburgh.

Mayer, M. M. (1961). In Kabat, E. A., "Kabat and Mayer's Experimental Immunochemistry", p. 149, Charles C. Thomas, Springfield, Illinois.

Murphy, T. M., and Mills, S. E. (1969). *J. Bact.*, **97**, 1310–1320.

Staub, A. M., and Forest, N. (1963). *Ann. Inst. Pasteur*, **104**, 71–83.

Steffen, D. L., Cocks, G. T., and Wilson, A. C. (1973). *J. Bact.*, **110**, 803–808.

Takatsy, G. (1956). *Acta Microbiol. Acad. Sci. Hung.*, **3**, 191–202.

Zabrieskie, J. B. (1964). In "Rheumatic Fever and Glomerulonephritis" (J. W. Uhr, Ed.), p. 53. Williams and Wilkins Co., Baltimore.

CHAPTER II

# The Identification of
# *Yersinia pseudotuberculosis*

ERNST THAL

*National Veterinary Institute, Stockholm, Sweden*

## I. INTRODUCTION

French and German writers have held divergent views as to who first detected *Yersinia pseudotuberculosis*. As early as 1883 Malassez and Vignal isolated coccobacilli from lesions in a guinea pig inoculated with pus from a child with tuberculosis meningitis. Their findings agree with a more detailed description of a Gram-negative coccoid bacterium by Pfeiffer in 1889 who gave it the name of *Bacillus pseudotuberculosis rodentium*.

The first direct isolation of *Y. pseudotuberculosis* from man was that of Saisawa (1913) from the blood of a patient with septicaemia in 1909. During the next four decades infections with *Y. pseudotuberculosis* have been reported sporadically from man and animals under a score of different names. For a list of the numerous synonyms the interested reader is referred to Index Bergeyana (1966).

In the early fifties Knapp and Masshoff (Knapp and Masshoff, 1954; Masshoff, 1953; Masshoff and Dölle, 1953) observed that *Y. pseudotuberculosis* relatively often caused a more or less benign mesenterial lymphadenitis. Since then, a reappraisal of the importance of this organism for disease in man has taken place and infections with *Y. pseudotuberculosis*

2

have been reported with increasing frequency from man and animals from most parts of the world.

The disease exists in a wide range of tame and wild animals (Hubbert *et al.*, 1971; Mair, 1973; Thal, 1954). In Europe wild mammals and birds constitute sylvatic reservoirs similar to those found in North America for the immunologically closely related *Y. pestis*, the cause of plague (see II). Animals kept and bred by man can transmit the infection. Outbreaks in farm animals, turkeys, mink as well as in laboratory animals, e.g. guinea-pigs, have been described. Afflicted cats, cage birds, e.g. parrots and canaries, can spread the disease to their surroundings (Fredriksson *et al.*, 1962; Pilet, 1955). In the enumeration of zoonotic diseases the World Health Organisation includes infections with *Y. pseudotuberculosis*.

Manifestations of infections with *Y. pseudotuberculosis* range from an acute fulminant form to chronic, or latent infection depending on the variable virulence of the invading strain and the intrinsic resistance of the host animal. In man symptoms of mesenterial lymphadenitis or gastro-enteritis are common (Daniels, 1973; Somov and Martinevsky, 1973). Epidemiologically, a preponderance of juvenile male patients and a higher seasonal incidence during spring and autumn have been observed (Knapp, 1959).

It is assumed that man is infected by direct or indirect contact (e.g. infected vegetables, berries).

## II. CLASSIFICATION

Since the beginning of the seventies international consensus has been reached for the name *Yersinia psuedotuberculosis*. By the resolutions of the International Committee on Systematic Bacteriology, *Pasteurella pseudotuberculosis*, *Pasteurella pestis* and *Pasteurella* X (syn. *Yersinia enterocolitica*) are to be separated from the genus *Pasteurella* to form the genus *Yersinia* (Mollaret and Knapp, 1972; Thal, 1974). In the eighth edition of Bergey's Manual of Determinative Bacteriology (Mollaret and Thal, 1974), the genus *Yersinia* comprises the three species *Y. pseudo-tuberculosis*, *Y. pestis* and *Y. enterocolitica* which have been incorporated into the family *Enterobacteriaceae*. As early as 1949 the generic name *Yersinia* was proposed by van Loghem for *Y. pseudotuberculosis* and *Y. pestis* in honour of Yersin, who discovered *Y. pestis* in 1884. By his proposal, van Loghem wished to emphasise the close immunological relationship between *Y. pseudotuberculosis* and *Y. pestis* (van Loghem, 1946). The inclusion of *Y. pseudotuberculosis* into the family *Enterobacteriaceae* was proposed in 1954 by Thal. In most respects these bacteria do not differ from other *Enterobacteriaceae*. The morphological and serological structure is similar. Bacteria of the genus *Yersinia* may be isolated from intestinal contents, several serogroups of *Y. pseudotuberculosis* share antigens with salmonellae (see III) and *Escherichia coli* (Mair and Fox, 1973). Numerical taxonomy (Smith and Thal, 1965), immunochemical structure (Brubaker *et al.*, 1973; Lüderitz *et al.*, 1966), overlapping phage sensitivity (Girard, 1942; Rische *et al.*, 1973) and transfer of genetic factors (Lawton *et al.*, 1968) all seem to justify the incorporation of the genus *Yersinia* into the family *Enterobacteriaceae*.

## III. IDENTIFICATION

### A. Staining and cultivation

Methods and media are as recommended for the identification of *Salmonella* (Edwards and Ewing, 1972; Kauffmann, 1966; Sedlak and Rische, 1968); *Y. pseudotuberculosis* is an aerobic, facultatively anaerobic organism when grown on ordinary liquid and solid media, growing as rods or coccoid forms, $0.5 \times 1.5$ $\mu$m, sometimes longer. All ordinary staining methods can be used. In most strains motility can be demonstrated in culture. Soft agar ("swarm agar") (0.4% agar) on plates or in U-tubes incubated at room temperature (*c.* 20°C) for at least 18 h should be used.

Motility and alkalinity (yellow) when grown on Leifson's desoxycholate citrate agar (DC-agar) and a positive urea-reaction are important presumptive differential characteristics as against *Y. pestis*. *Y. pestis* is non-motile, grows as red colonies on DC-agar and gives a negative urea reaction (Thal and Chen, 1955).

For the cultivation of *Y. pseudotuberculosis*, the optimal temperature is 30°C. *Y. pseudotuberculosis* usually grows slower than *Salmonella* for example and only after some 48 h is a colony size of 1–2 mm attained. Some strains, especially in O-group III (see III C), exhibit flat, rough colonies which nevertheless turn out to be serologically S-forms. Strains with a tendency to show spontaneous agglutination may often be transformed into smooth variants by 2–4 selective passages on swarm-agar. After several days, the colonies may have mucoid wall-like borders.

Passage of samples through different *Salmonella*-enrichment broth tubes (e.g. tetrathionate and selenite broths) suppresses the growth, and hampers the isolation, of *Y. pseudotuberculosis*. The search for *Y. pseudotuberculosis* should therefore be attempted by direct plating of faeces. Two tenfold dilutions should be plated.

### B. Biochemical characteristics

Biochemical reactions of more than 1500 *Y. pseudotuberculosis* strains of different serological groups from all over the world have proved to be homogeneous during a period of some 25 years. Negligible deviations in a few reactions have been noted. A selection of some useful tests which should suffice for routine diagnostic work is presented in Table I. For a more complete scheme, the reader is referred to Bergey's 8th edition (Mollaret and Thal, 1974).

TABLE I

**Biochemical reaction helpful in the characterisation of *Yersinia pseudo-tuberculosis***

| | |
|---|---|
| Citrate | − |
| Esculin | + |
| $\beta$-Galactosidase | + |
| Gas from glucose | − |
| Indole | − |
| KCN | − |
| MR reaction | + |
| VP | − |
| Nitrate | + |
| Ornithine decarboxylase | − |
| Oxidase | − |
| Rhamnose | + |
| Salicin | + |
| Sorbitol | − |
| Sucrose | − |
| Xylose | + |
| Urease | + |

## C. Antigenic analysis

If the biochemical reactions fit in with the above mentioned characteristics of *Y. pseudotuberculosis* and serological typing seems of interest, strains can be sent to one of the reference laboratories (see V) especially interested in *Y. pseudotuberculosis*. Based upon the early work by Schütze (1928) and Kauffmann (1932), an antigenic scheme of *Y. pseudotuber-culosis* regarding O-, H- and R-antigens has been developed during the last decades (Thal and Knapp, 1971). Although the species is homogeneous in its biochemical reactions, it is serologically divided into 6 distinct, not overlapping O-groups. Some strains may in addition contain an immunologically homogeneous toxic component. They are to be considered as exotoxins which can be neutralised by antisera or protected against by the administration of toxoid (Brown *et al.*, 1969; Schär and Thal, 1955).

The mechanisms of immunity for infectiousness and toxicity are different and are presented in Table II.

## TABLE II

Anti-infectious and antitoxic immunity in *Yersinia pseudotuberculosis*

| | Protection against | | |
|---|---|---|---|
| Immunity | Infection | Toxin | Toxi-infection |
| Anti-infectious (with living bacteria) | + | − | − |
| Antitoxic (toxoid) | − | + | + |

Serological relationships with *Y. pestis* (R-antigen-complex) exist and are illustrated in Table III (Bhatnagar, 1940; Thal, 1956 and 1973).

The revised antigenic scheme (Table IV) published in 1971 is still valid as, to our knowledge, no strains deviating from it have yet been isolated (to May, 1977).

In ordinary diagnostic laboratories O-grouping seems in the author's opinion to be sufficient. Laboratories should have a set of 6 group strains and the homologous antisera available. If further analysis is required for epidemiological purposes, the definition of the sub-groups and H-antigens should be performed by central laboratories, or by one of the special laboratories listed under V.

## TABLE III

Crossimmunity in *Yersinia pestis* and *Yersinia pseudotuberculosis* (demonstrated in guinea pigs)

| | Infection with | |
|---|---|---|
| Immunisation with | *Y. pestis* 195/P | *Y. pseudotuberculosis* 14/1 |
| *Y. pestis* EV 76 (living bacteria) | 10/10† | 2/10 |
| *Y. pseudotuberculosis* 32/IV (living bacteria) | 10/10 | 9/10 |
| Controls | 0/10 | 0/10 |

† Numerator: surviving animals. Denominator: number of infected animals, e.g. 2/10: 2 guinea pigs of 10 infected survived.

## TABLE IV

### The antigenic scheme of *Yersinia pseudotuberculosis*

| O-groups | O-subgroups | Rough antigens "R" | O-antigens (thermostable) | H-antigens (thermolabile) |
|---|---|---|---|---|
| I | A | (1) | 2, 3 | a, c |
|   | B | (1) | 2, 4 | a, c |
| II | A | (1) | 5, 6 | a, d |
|   | B | (1) | 5, 7 | a, d |
| III |  | (1) | 8 | a |
| IV | A | (1) | 9, 11 | b; a, b |
|   | B | (1) | 9, 12 | a, b, d |
| V | A | (1) | 10, 14 | a; a, e (b) |
|   | B | (1) | 10, 15 | a |
| VI |  | (1) | 13 | a |

In Europe, serogroup I has been most prevalent (Thal, 1954), although groups II and III are not unusual. Strains of groups IV and V have been rare and group VI has only been found in Japan.

Kauffmann (1932) identified common antigens between group II and *Salmonella* group B. Knapp (1959) proved a serological relationship between group D strains of *Salmonella* and *Y. pseudotuberculosis* group IV. The common immunochemical components are abequose for *Salmonella* group B and *Y. pseudotuberculosis* group II and tyvelose for *Salmonella* group D and *Y. pseudotuberculosis* group IV (Brubaker *et al.*, 1973; Lüderitz, 1966).

Immunofluorescence has been of limited use for diagnosing *Y. pseudotuberculosis*. Cross-reactions with *Salmonella* persist up to high dilutions and, as with *Salmonella*, there are difficulties on account of cross-reactions with other *Enterobacteriaceae*. At this stage, the author sees no advantages of this method in routine diagnostic work. For the technique and analysis of immunofluorescence, the reader is referred to the work by K.-A. Karlsson (1975).

## D.  The preparation of antigen for agglutination of H- and O-antigens and for the production of diagnostic rabbit sera

Cultures of the different group strains (I-VI) are grown in sugar-free agar (about 1·5%) for 48 h and harvested with saline. The turbidity of the suspension is adjusted to Wellcome 4 for the production of diagnostic sera and to Wellcome 2 for the agglutination of antigen.

To produce *O-antigen*, the bacterial suspensions are boiled for 2·5 h and tested for sterility and for their ability to be agglutinated.

Rabbits are injected intravenously in increasing doses (0·25, 0·5, 1·0, 1·5 and 2 ml) at 3 to 4 days intervals. About 1 week after the last injection, the rabbits are bled, if the homologous agglutination titre is satisfactory, e.g. from 1/320 – upwards. The serum is inactivated in a water bath at 56°C for 30 min and after addition of merthiolate (1 : 10,000) stored in the refrigerator.

For agglutination, the antigen diluted to Wellcome 2 needs to be boiled only for half an hour. The agglutination is performed as with other *Enterobacteriaceae*. The author uses 0·5 ml of the serum dilution together with 5 ml antigen in every tube starting with 1 : 10. The results are read after 20 h incubation at 37°C. The last tube with a clear supernatant is recorded as the serum titre.

Cultures for the preparation of *H-antigen* should show good motility after growth on soft agar (0·4%) at 22°C. If necessary, the motility may be enhanced by several passages. As soon as satisfactory "swarming" of the culture has been achieved, the culture is harvested with physiological saline to which 0·3% formalin is added. After incubation overnight at 37°C and 3 days preservation in the refrigerator, the suspension is tested for sterility and specific agglutination.

For the injection of rabbits the turbidity of the H-antigen is adjusted to Wellcome 4, whereas for the agglutination antigen a turbidity of Wellcome 2 is used. The rabbits are injected and treated according to the same schedule as for the O-antigen. The rabbits are bled when a titre of at least 1 : 600 has been reached.

H-agglutination tests are started at a serum dilution of 1 : 100 and read after 4 h incubation at 37°C in a water bath.


## IV. ANIMAL INOCULATION

With regard to their effect on laboratory animals, three different kinds of *Y. pseudotuberculosis* strains can be discerned; invading strains, invading and toxic strains and avirulent strains. Avirulent strains have to be identified by biochemical and serological tests, whereas for virulent strains in the isolation of a pure culture the *guinea-pig* is most suitable. To determine toxicity, however, additional tests with filtrates in *mice* are recommended. In guinea-pigs, death usually occurs within 7 to 21 days after the subcutaneous injection of 0·5 ml of the suspected material diluted so that it will easily pass through the injection needle. At autopsy, miliary necroses are found in the parenchymatous organs (liver, spleen, kidney, bone marrow). These are macroscopically similar to necroses produced by other infectious diseases as, for example, salmonellosis and listeriosis, but pathologists consider the clusters of bacteria seen within the necrotic foci as characteristic of *Y. pseudotuberculosis* (Borg *et al.*, 1969).

As mentioned above, *mice* are the animals of choice for work with the *Y. pseudotuberculosis* toxins (Schildt *et al.*, 1964–1965).

*Rabbits.* Experimentally infected rabbits can develop an acute or chronic infection or merely show a specific agglutination titre or also succumb to the toxic component of a virulent strain. *Y. pseudotuberculosis* toxin causes necrotic lesions in the skin test in rabbits which can be neutralised with antisera. Also ligated rabbit intestines have proved to be a useful experimental system for the demonstration of the toxin of *Y. pseudotuberculosis* (Söderlind and Thal, 1975).

*Rats* seem to be resistant to infection with *Y. pseudotuberculosis*. This has been considered a differential characteristic against *Y. pestis* to which rats respond. However, rats may succumb to the toxin of *Y. pseudotuberculosis* (Schär and Thal, 1955).

# V. INTERNATIONAL REFERENCE LABORATORIES IN EUROPE

| | |
|---|---|
| *France* | Institut Pasteur |
| | Service de la peste |
| | *Prof. H. H. Mollaret* |
| | Rue du Docteur Roux |
| | Paris XIII |
| *West Germany* | University Nürnberg-Erlangen |
| | Institut für Hygiene und Med. Mikrobiologie |
| | *Prof. W. Knapp* |
| | Wasserturmstrasse 3 |
| | 8520 Erlangen |
| *United Kingdom* | Public Health Laboratory Service |
| | *Director – Dr. N. S. Mair* |
| | Groby Road Hospital |
| | Groby Road |
| | Leicester LE3 9QE |

## REFERENCES

Bhatnagar, S. S. (1940). *Indian J. med. Res.*, **28**, 1–17.

Borg, K., Hanko, E., Krunagević, T., Nilsson, N. G., and Nilsson, P. O. (1969). *Nordisk vet.-med.*, **21**, 95–104.

Brown, J. A., West, W. L., Banks, W. M., and Marshall, J. D. (1969). *J. infec. Dis.*, **119**, 229–236.

Brubaker, R. R., Hellerqvist, C. G., Lindberg, B., and Samuelsson, K. (1973). "Contributions to Microbiology and Immunology", **2**, 6–9. Karger, Basel.

Daniels, J. J. H. M. (1973). "Contributions to Microbiology and Immunology", **2**, 210–213. Karger, Basel.

Edwards, P. R., and Ewing, W. H. (1972). Identification of *Enterobacteriaceae*. 3rd edn. Burgess Publ. Co., Minneapolis, Minnesota.

Fredriksson, W., Kiaer, W., and Lauridsen, L. (1962). *Acta path. microbiol. scand. Suppl.*, **154**, 277.

Girard, G. (1942). *Ann. Inst. Pasteur*, **68**, 476–478.

Hubbert, W. T., Petenyl, C. W., Glasgow, L. A. *et al.* (1971). *Am. J. Trop. Med. Hyg.*, **20**, 679–684.

Index Bergeyana (1966). The Williams and Wilkins Co., Baltimore.

Karlsson, K.-A. (1975). Thesis, Stockholm.

Kauffmann, F. (1932). *Z. Hyg. Infekt.-Kr.*, **114**, 97–105.

Kauffmann, F. (1966). The Bacteriology of *Enterobacteriaceae*. Munksgaard, Copenhagen.

Knapp, W. (1959). *Ergebn. Mikrobiol. Immunforsch. exp. Ther.*, **32**, 196–269.

Knapp, W., and Masshoff, W. (1954). *Dtsch. med. Wschr.*, 1266.

Knapp, W., and Lebek, G. (1967). *Schweiz Z. Path. Microbiol.*, **30**, 103.

Lawton, W. D., Morris, B. C., and Burrows, T. W. (1968). *Symp. series Immunobiol. Standard. Symp. on Pseudotuberculosis*, **9**, 285–393. Karger, Basel.

Loghem van, J. J. (1946). *Ann. Inst. Pasteur*, **72**, 975.

Lüderitz, O., Staub, A. M., and Westphal, O. (1966). *Bact. Rev.*, **30**, 192–255.

Mair, N. S. (1973). *J. Wildlife Disease*, **9**, 64–71.

Mair, N. S., and Fox, E. (1973). "Contributions to Microbiology and Immunology" **2**, 180–183.

Malassez, L., and Vignal, W. (1883). *Arch. Physiol. norm. path.*, **2**, 369–412.

Masshoff, W. (1953). *Dtsch. med. Wschr.*, 532.

Masshoff, W., and Dölle, W. (1953). *Virchows Arch. path. Anat.*, **323**, 664.

Mollaret, H. H., and Knapp, W. (1972). *Intern. J. Systematic Bact.*, **22**, 401.

Mollaret, H. H., and Thal, E. (1974). "Genus *Yersinia*". Bergey's Manual of Determinative Bact. 8th ed. The Williams and Wilkins Co., Baltimore.

Pfeiffer, A. (1889). "Ueber die bacilläre Pseudotuberculose bei Nagetieren", Leipzig.

Pilet, C. (1955). "La pseudotuberculose du chat". Thesis, Lyon.

Rische, H., Beer, W., Seltmann, G., Thal, E., and Horn, G. (1973). "Contributions to Microbiology and Immunology", **2**, 23–26. Karger, Basel.

Saisawa, K. (1913). *Z. Hyg. Infekt.-Kr.*, **73**, 353–400.

Schär, M., and Thal, E. (1955). *Proc. Soc. exp. Biol. Med.*, **88**, 39–42.

Schildt, B., Karlsson, K.-A., and Thal, E. (1968). Intermedes Proceedings, Research Institute of National Defence, Stockholm, Sweden. Combined Injuries and Shock, pp. 75–81. No. 36.

Schütze, H. (1928). *Arch. Hyg.*, **100**, 181–194.

Sedlak, J., and Rische, H. (1968). VEB Georg Thieme, Leipzig. Enterobacteriaceae-infektionen.

Smith, J., and Thal, E. (1965). *Acta path. microbiol. scand.*, **64**, 213–223.

Somov, G. P., and Martinevsky, I. L. (1973). "Contributions to Microbiology and Immunology", **2**, 214–216. Karger, Basel.

Söderlind, O., and Thal, E. (1975). 6th Int. Colloquium on phage typing and other laboratory methods for epidemiological surveillance. Wernigerode 1975. Abstracts, 45.

Thal, E. (1954). "Pasteurella pseudotuberculosis". Thesis, Stockholm.

Thal, E. (1956). *Ann. Inst. Pasteur*, **91**, 68–74.

Thal, E. (1973). "Contributions to Microbiology and Immunology", **2**, 190–195. Karger, Basel.

Thal, E. (1974). *Berl. Munch. Tierarztl. Wschr.*, **87**, 212–214.

Thal, E., and Chen, T. H. (1955). *J. Bact.*, **69**, 103–104.

Thal, E., and Knapp, W. (1971). *Symp. series Immunobiol. Standard.*, **15**, 219–222. Karger, Basel.

CHAPTER III

# Bacteriophage Typing of
# *Yersinia enterocolitica*

TOM BERGAN

*Department of Microbiology, Institute of Pharmacy, University of Oslo*

## I. INTRODUCTION

Serological characterisation of *Yersinia enterocolitica* is the epidemiological typing method which has gained the widest acceptance. No system for bacteriocine typing has been worked out and, indeed, there have been reports to the effect that this species does not have bacteriocins (Nicolle *et al.*, 1967). In this Chapter the available systems for phage typing of *Y. enterocolitica* will be presented.

## A. Presence of phages in *Y. enterocolitica*

A considerable proportion of isolates of *Y. enterocolitica* are lysogenic. The frequencies vary, but on the average 70–85% of strains may harbour phages (Nicolle *et al.*, 1967; Nicolle *et al.*, 1968; Niléhn and Ericson, 1969). In one instance, 60% was seen primarily and 67% upon induction with ultraviolet (UV) light (Mollaret and Nicolle, 1965). The methods used have involved bacteria-free lysates spotted on to a number of sensitive strains. Occasionally several phages have been isolated from the same culture. Nicolle *et al.* (1969) found that the most frequent lysotype in their system, lysotype III, was lysogenic at a frequency of 99%. On the other hand phages have never been recovered from lysotype IXa.

## B. Basis for phage typing

The *Y. enterocolitica* phages differ in their host ranges, such that different patterns of sensitivity are obtained with a given set of phages on different bacterial isolates (Nicolle *et al.*, 1968; Niléhn 1973). Host modifications of phage spectra are easily obtained on propagation in different susceptible strains (Niléhn, 1969). Since natural variation in host sensitivity is considerable, the phage typing sets have been mostly composed of wild phages, but phages changed by host modification have also been included (Nicolle, 1973; Niléhn, 1973).

## C. Properties of *Y. enterocolitica* phages

The bacteriophages are differentiable serologically, but this observation has received little attention (Niléhn, 1973).

Temperature sensitivity has been studied on phage 1A which is inactivated at 65°C after 30 min, but is unaffected by 50°C when 25°C is used as control (Niléhn, 1973).

Propagation of phages in *Y. enterocolitica* does not take place at 37°C, but occurs readily at 22°C. Absorption is assumed to take place at 37°C, however, and results in growth inhibition (Niléhn and Ericson, 1969).

## II. PHAGE TYPING SETS

Two phage typing sets have been worked out for *Y. enterocolitica*, one in France (Nicolle, 1973), and one in Sweden (Niléhn, 1973).

## A. French phage typing set

The first work on phage typing of *Y. enterocolitica* was carried out by Nicolle *et al.* (1967) and has subsequently been refined and studied in a series of publications (Nicolle *et al.*, 1968; Nicolle *et al.*, 1969; Nicolle *et al.*, 1973; Nicolle, 1973). In the first version, the typing set consisted of ten phages (Table I). These have been termed 1–10 by Arabic numerals at first, and later substituted by I–X in Roman designations (Table II). A subset has been developed to type the non-typable strains which after the primary typing were designated X, the last of the 10 lysotypes recognised (Table III) (Nicolle *et al.*, 1973). The phages of the primary set were obtained from lysogenic strains, whereas the secondary set was derived from

## TABLE I

**Lytic types for phage typing set of *Yersinia enterocolitica* (Nicolle *et al.,* 1968)**

| Types | Phages | | | | | | | | | |
|-------|----|-----|----|----|----|----|----|-----|----|----|
|       | 1  | 2   | 3  | 4  | 5  | 6  | 7  | 8   | 9  | 10 |
| 1     | ±  | ±   | LC | ±  | −  | ±  | LC | LC  | −  | LC |
| 2     | ±  | ±   | LC | LC | ±  | LC | ±  | LC  | −  | LC |
| 3     | LC | LCV | LC | −  | ±  | LC | ±  | LC  | ±  | LC |
| 4     | LC | −   | LC | LC | LC | −  | LC | LC  | −  | LC |
| 5     | LC | −   | LC | −  | LC | −  | −  | LC  | −  | −  |
| 6     | −  | −   | −  | −  | −  | −  | −  | +++ | −  | −  |
| 7     | LC | −   | LC | −  | LC | LC | LC | LC  | −  | LC |
| 8     | LC | LC  | LC | LC | LC | LC | LC | −   | ±  | LC |
| 9     | LC | LC  | LC | LC | LC | LC | LC | LC  | LC | LC |
| 10    | −  | −   | −  | −  | −  | −  | −  | −   | −  | −  |

LC = lytic confluence
LCV = lytic confluence disturbed by resistant colonies in the lytic zone
+ + + = numerous plaques
± = variable lytic activity
− = no lysis

## TABLE II

**Lytic types for phage typing set of *Yersinia enterocolitica* (Nicolle, 1973)**

| Phage types or groups | Aa | b | c | d | e | f | g | h | i | j | k | l |
|-----------------------|----|----|----|----|----|----|----|-----|----|----|----|----|
| I    | ± | ± | LC | ± | − | ± | LC | LC | − | LC | | |
| II   | ± | ± | LC | LC | ± | LC | ± | LC | − | LC | | |
| III  | LC | LC | LC | − | ± | LC | ± | LC | ± | LC | | |
| IV   | LC | − | LC | LC | LC | − | LC | LC | − | LC | | |
| V    | LC | − | LC | − | LC | − | − | LC | − | − | | |
| VI   | − | − | − | − | − | − | − | +++ | − | − | | |
| VII  | LC | − | LC | − | LC | LC | LC | LC | − | LC | | |
| VIII | LC | LC | LC | LC | LC | LC | LC | − | LC | LC | LC | − |
| IXa  | LC | LC | LC | LC | LC | LC | LC | LC | LC | LC | LC | LC |
| IXb  | LC | LC | LC | LC | LC | LC | LC | LC | LC | LC | − | LC |
| IXc  | − | LC | LC | LC | LC | LC | − | LC | LC | LC | − | LC |
| Group X  | − | − | − | − | − | − | − | − | − | − | − | − |
| Group XI | Lytic pictures that cannot be identified as above type | | | | | | | | | | | |

Symbols as in Table I.

T. BERGAN

## TABLE III

**Lytic types of phage typing set modified by addition of subset active in** *Yerisinia enterocoliticia* (Nicolle *et al.,* 1973)

| Sub groups | 1 | 2 | 3 | 4 | 5 | 6 | 7 | 8 | 9 | 10 | 11 | 12 |
|---|---|---|---|---|---|---|---|---|---|---|---|---|
| 1 | − | LC | LC | LC | LC | + | LC | LC | LC | − | LC | +++ |
| 2 | − | LC | LC | LC | LC | − | LC | ++ | LC | − | − | − |
| 3 | − | − | LC | LC | − | − | − | − | LC | LC | LC | LC |
| 4 | − | − | LC | − | − | − | − | − | LC | − | LC | LC |
| 5 | − | − | − | +++ | − | − | − | − | LC | >LC | − | +++ |
| 6 | − | − | LC | LC | LC | − | +++ | − | − | − | − | − |
| 7 | − | − | LC | − | − | − | − | − | LC | − | − | − |
| 8 | − | − | LC | − | − | − | − | − | − | − | − | − |
| 9 | − | − | − | LC | − | − | − | − | − | − | − | − |
| 10 | − | − | − | − | − | − | − | − | − | − | − | − |

CL = confluent lysis
± = variable lytic activity
− = no lysis

sewage water. Most recently, the primary set has been extended by another two phages which are used principally to subdivide the lysotype IX into three subentities, $IX_a$, $IX_b$, and lysogroup $IX_z$ (Table II). The phages are now designated by small letters a–l with the two last entries named k and l. Lytic codes that cannot be identified by the key in Table II are lumped together in one group XI (Nicolle, 1973).

The methodology regarding procedures for propagation and media has not been described in detail but the procedures generally employed in enteric phage typing would be applicable (Anderson and Williams, 1956) with the modification that an incubation temperature of 25°C is necessary for development of *Y. enterocolitica* plaques.

## B. Swedish phage typing set

Another phage typing set has been developed by Niléhn (1969) and Niléhn and Ericson (1969). This consists of the 7 *Y. enterocolitica* phages B1, B2, A1, E1, A2, C31, and C61, and three different preparations of the *Y. pseudotuberculosis* phage Girard 97/II which has been grown in three different systems:

(a) on *Y. pseudotuberculosis* PS97/II at 25°C
(b) on *Y. pseudotuberculosis* PS97/II at 37°C
(c) on *Y. enterocolitica* P311 at 37°C

Thus 10 phage preparations are used, 3 of which are host modifications of the same *Y. pseudotuberculosis* phage. The *Y. enterocolitica* phages were all derived from 5 phages, A–E; in the case of C31 and C61 two new propagating strains were selected.

A set of 12 strains of *Y. enterocolitica* is used to test the lytic spectra of the phages. The designations of these indicator strains and the lytic spectra are shown in Table IV. The Table also provides a key to type designations according to different lytic patterns totalling 12 different types.

TABLE IV

**Lytic spectrum for phage typing of *Yersinia enterocolitica* (Niléhn, 1973)**

| *Y.* phage pattern 25°C | Type strain | *Y. enterocolitica* phages | | | | | | | PST phage | |
|---|---|---|---|---|---|---|---|---|---|---|
| | | B1 | B2 | A1 | E1 | A2 | C31 | C61 | PST/97 | PST/311 |
| Ia | T 347 | L | L | L | L | L | L | L | | |
| Ib | P 131 | (−) | L | L | L | L | L | L | | |
| Ic | P 420 | | | L | L | L | L | L | | |
| II | M Y1 | L | L | | L | L | L | L | | |
| III | MY 75 | L | L | | | L | L | L | | |
| IVa | P 77 | | | | | | L | L | | |
| IVb | MY 31 | | | | | | L | | | |
| Va | P 248 | L | L | | | L | | (−) | | |
| Vb | T 338 | L | L | | | | | | | |
| VI | PS 97 | | | | | | | L | L | L |
| VII | P 76 | | | | | | | | | |
| VIII | P 311 | | | | | | | | | |
| **37°C** | | | | | | | | | | |
| VI | PS 97 | | | | | | | L | L | L |
| VII | P 76 | | | | | | | L | L | L |
| VIII | P 311 | | | | | | | | (−) | L |

L = > 50 Plaques with RTD; (−) = weak reaction with RTD.

In this system, the phages are propagated by the soft agar technique according to Adams (1959). Typing proceeds at both 25° and 37°C. Details of media and technique are given in the Appendix.

At RTD the action of all phages except C61 is restricted to the species *Y. enterocolitica*. Phage C61 also lyses strains of *Y. pestis* and *Y. pseudotuberculosis* at both 25° and 37°C. At concentrations higher than the RTD, other phages are also active against *Y. pseudotuberculosis* at 37°C. The *Y. pseudotuberculosis* phage is lytic on *Y. enterocolitica* strains also at 37°C – in addition to the activity at 25°C which is usual in this species.

The phage set only yields a poor subdivision of the isolates. Of 295 human isolates, two-thirds were phage type II. The other type with any degree of distribution is number III. All but 14 strains of 295 human isolates were distributed among these two types (Niléhn, 1973). In strains from animals, two types, Ic and III constituted some 50% of the isolates (Niléhn, 1973). Few strains have proved non-typable. This homogeneity of type distribution parallels completely the situation for serogrouping. For instance, all but 10 strains of 342 isolates belonged to O-types 3 and 9. O-serotype 3 represents 85% of the isolates (Niléhn, 1973).

## C. Interpretation of results

Nicolle *et al.* (1973) distinguished 11 types and one non-typable group with their typing set. Niléhn (1973) with her set distinguished 12 lysotypes. The last two types are identified by their reaction to the *Y. pseudotuberculosis* phages and by incubation at 37°C in addition to 25°C.

## D. Assessment of sets

Both the French and the Swedish typing sets have been tested on a large number of isolates, predominantly of human origin. It is evident that both sets succeed also in typing the non-human strains adequately. Nicolle *et al.* (1973) have reported the typing of 1252 strains and Niléhn (1973), with the Swedish set of 352 strains (Table V). In the former case a large portion was of animal origin, in the latter case one-fifth of the strains came from animals. Approximately one-fifth of the French typed strains were non-typable with the primary set, but appear to have been typed by the secondary set developed later. The Swedish set, on the other hand, has only left 1% of the strains untyped. Both typing sets result in what would seem to be an unsatisfactory separation in to different lysotypes. One type dominated in each case. Whereas human strains are divided mainly among two phage types in the Niléhn set (1973), one type is completely dominating in the French alternative. Of 839 strains typed – derived from various animals in addition to man – phage type VIII constituted 538 strains and 135 were non-typable Nicolle (1973). Among the human strains type VIII was so dominating that only 36 strains of 658 were of other types, and only three other types were represented among these; indeed, of the non-VIII strains, 34 of the 36 strains belonged to phage type IX. In the total material a large proportion of the strains from chinchilla came from one single phage type. Even after application of the secondary typing set to all the non-typable ones from the first set, only one more phage type dominates the material and thus adds little useful information.

Accordingly, neither set is completely satisfactory when it comes to producing a number of distinct epidemic types. When animal strains are typed, either set reveals mostly one dominating type among the strains derived from each species. This is in spite of the fact that the strains in both instances have been obtained from world wide distributed sources. In some instances, there have been geographic distinctions between the strains from a given animal species, but on the whole homogeneity of types is the general impression. Since the number of types studied is so small, there is little epidemiological information to gather from the results.

It may also be that more valuable information would be obtained if the phage sets used were selected with the aid of numerical taxonomic analysis of their lytic spectra as has been done successfully with typing phages of *Pseudomonas aeruginosa* (Bergan, 1972).

## III. EPIDEMIOLOGY

*Human strains* have belonged to the French lysotypes VIII and IX (Nicolle *et al.*, 1968; Nicolle, 1973). Approximately half of the primarily non-typable strains belonged to the same phage type as determined by the secondary typing set.

The Swedish phage typing set succeded in subdividing the O-serotype 3 which dominated among human strains into lysotypes I, II, III, and IV (Niléhn, 1969). Type II has dominated in Sweden, whereas III has been obtained frequently from other countries.

In five instances, *Y. enterocolitica* has been isolated simultaneously from several family members with symptoms of disease (Niléhn, 1969). Within the same household, strains of the same type were usually found. Only in one instance were two types, II and III, isolated in parallel from two individuals within the same family. The primary source of the infection in this case was not determined.

*Pig strains.* Curiously enough, the dominating type in pigs has been the lysotype III by the Niléhn (1973) set. This happens to be the same as that dominating among non-Swedish human isolates. A similar pattern was observed with the French set, where type VIII was frequent.

*Monkey strains.* Strains from monkey have shown patterns similar to the human strains.

*Hare strains. Y. enterocolitica* from hare is distinguished by being more evenly distributed among more lysotypes.

*Chinchilla strains* are also mainly one type, but several types are represented.

*Cow and horse strains* tend to show the same type distribution as found in man. One might speculate on the possibility that this concordance in types between man and his major domestic animals has epidemiological implications.

On the basis of the cross-typings with the two typing sets on the various animal strains, it would appear that rough estimates of how French types may be translated into the Swedish types are warranted in a few instances:

| French types | Scandinavian type |
|:---:|:---:|
| II | Ib |
| VII9 | III |
| IX | II |

## IV. REPRODUCIBILITY

The constancy of types has not been systematically studied for *Y. enterocolitica*. One may speculate that variability must be small since so few types are identified for each animal species. However, lysogenic conversion has been shown to occur in *Y. enterocolitica* (Nicolle *et al.*, 1969). For instance, strains of the French typing set lysotype IX which is attacked easily by all typing phages are converted into other types fairly easily by lysogenisation. In one instance, a type IX *Y. enterocolitica* was converted to a type II by phage obtained from a type II bacterial strain. In parallel with the phage receptor changes, it would be likely that bacteriophages may also contribute to the determination of serological characteristics of *Y. enterocolitica* as has been the case in lysogenic conversion of *Salmonella* (Escobar and Edwards, 1968; Nicolle *et al.*, 1962; Robbins and Uchida, 1962) and *Pseudomonas aeruginosa* (Bergan and Midtvedt, 1975). In a species with such a high frequency of lysogenisation as *Y. enterocolitica*, one might expect some instability in phage type. On the other hand, since the number of different lysotypes distinguished is low, it may be that variability is low. Only specifically planned studies directed toward the question of reproducibility will give a definite answer to the question.

## V. RELATIONSHIP BETWEEN LYSOTYPE AND SEROGROUP SPECIFICITY

The strains of O-serotype 3 which is the dominating serogroup in man is subdivided by phage typing into 6 different phage groups, although 2 of them for all practical purposes comprise the whole group (Niléhn, 1973). On the other hand, all O-serotype 9 and most O-serotype 6, 7, and 8 and all serologically non-groupable strains have been non-typable by phage typing.

In the French system, a lysotype obtained by subdividing lysotype X with a secondary phage set, $X_3$, has been responsible for all 0 : 9 serogroup strains (Nicolle, 1973).

## VI. TYPABILITY OF NON-ENTEROCOLITICA

Strains of *Y. pestis* and *Y. pseudotuberculosis* have been susceptible to some phages in the typing set of Niléhn (1973). Indeed, three variants of the same phage strain in the Swedish typing set have been derived from *Y. pseudotuberculosis*.

Niléhn (1973) noted a close relationship between *Y. enterocolitica* and *Y. pseudotuberculosis* from the viewpoint of susceptibility to the Swedish phage typing set. The taxonomic implications of these observations are obvious. Indeed, one of the new phage types was represented by a single strain, the taxonomic position of which could not be fully determined. Other members of the *Enterobacteriaceae* have proved resistant to the phages regardless of the incubation temperature employed. For this reason a particularly avid phage, C31, has been put forward as a useful taxonomic tool. Used at 10 or 100 times the RTD on the host propagating strain, C31 may be used in routine laboratory diagnostics as a species specific test for *Y. enterocolitica* (Niléhn, 1973).

# VII. APPENDIX

## A. Phage propagation

After purity of a phage has been determined by single plaque morphology with the overlayer method of Adams (1958), a phage may be propagated. A plate is then prepared where a heavy inoculum of phage on its host propagating strain results in confluent lysis. After lysis has occurred, the phage preparation is harvested into 1 ml of broth. This is centrifuged and the supernatant separated. Of this, 0·3 ml is mixed with 0·3 ml of an overnight broth culture of the host propagating strain grown at 22°C and transferred to 30 ml of Hartley's broth as prepared by Niléhn and Ericson (1969). The culture is incubated at 22°C for 6 h and then transferred to 4°C over-night.

The next morning, the supernatant after centrifugation is separated and filtered through a 0·22 $\mu$m membrane filter.

## B. Preservation of phages and host propagating strains

The phages are kept at 4°C and the bacterial strains freeze dried.

## C. Determination of routine test dilution (RTD)

The day before typing, the RTD is determined by ten-fold dilutions of the phage stock. For this a 6 h culture of the host strain grown at 22°C and plated on a lawn is used. After drying the bacterial inoculum, spots of the phage dilutions are applied to the plate surface. The seed strain has been grown in meat extract broth (Bacto Beef Extract (Difco) 0·5% (w/v)

NaCl 0·3% (w/v), Na₂HPO₄.2H₂O 0·2% (w/v), pH 7·6–7·8. Autoclaving at 120°C for 30 min).

TABLE V

**Distribution of phage types among various hosts to *Yersinia enterocolitica* (Niléhn, 1973)**

| Phage type | Animal species | Man |
|------------|----------------|-----|
| Ia | Chinchilla, hare | + |
| Ib | Chinchilla, hare | + |
| Ic | Hare | + |
| II | Dog | + |
| III | Pig, monkey, guinea pig | + |
| IVa | Chinchilla | − |
| IVb | Pig | + |
| Va | Chinchilla | − |
| Vb | Chinchilla | − |
| VI | None* | − |
| VII | Chinchilla | − |
| VIII | Pig, hare | + |

+ = present
− = absent
* = Phage type VI is *Y. pseudotuberculosis*

The result is read after the plate has been incubated for one and two days. The dilution showing near confluent lysis is used as the RTD.

## D. Determination of lytic spectrum

The lytic spectra of the phages are determined by application of RTD on all propagating strains (Table V). In this instance, preincubation of seeded plates is carried out for 6 h at 22°C before the application of the phage suspensions.

The results are graded as follows:

++    more than 50 plaques
+    20 to 50 plaques
±    less than 20 plaques
−    no plaque formation

For the full lytic spectrum, the phages are applied at different ten-fold dilutions on all propagating strains and the results recorded according to the concentration of phage compared to its homologous RTD which produces a near confluent lysis on the various strains:

5 a reaction identical to the homologous one at RTD

4 near confluent (NC) lysis produced by a phage suspension which is $10-10^2$ more concentrated than RTD

3 NC lysis produced by a phage suspension which is $10^3-10^4$ times more concentrated than the RTD

2 NC lysis produced by a phage suspension which is $10^5-10^6$ times more concentrated than RTD

1 very weak lysis at concentrations above $10^6$

0 No lysis

The medium used for determination of the spectrum and for phage typing has the following composition:

| | |
|---|---|
| Beef-Extract Lab-Lemco (Oxoid) | 75 g |
| Peptone Orthana Special (Orthana) | 125 g |
| NaCl | 50 g |
| Bacto Agar (Difco) | 90 g |
| Distilled water to | 1000 ml |

The pH is adjusted to 7·4 and the medium is autoclaved at 120°C for 30 min. After autoclaving, sterilised $CaCl_2$ is added to a final concentration of 200 $\mu$g/ml. This must be done only after the medium has been cooled to avoid precipitation.

### E. Phage typing procedure

The strains to be tested are first preincubated at 22°C in meat extract broth for 6 h. The suspension, then obtained, is used to inoculate the plates. This will yield a dense, confluent growth of the bacteria. After drying the plates, the phages are applied at a concentration corresponding to RTD.

The plates are read after one and two days incubation at 25°C. A hand lens should be used for inspection of the plates. All typing must be done in duplicate.

### REFERENCES

Adams, M. H. (1958). "Bacteriophages". Interscience Publishers, New York.
Anderson, E. S. and Williams, R. E. O. (1956). *J. clin. Path.*, **9**, 94–127.
Bergan, T. (1972). *Acta path. microbiol. scand. Sect. B*, **80**, 189–201.
Bergan, T. and Midtvedt, T. (1975). *Acta path. microbiol. scand. Sect. B*, **83**, 1–9.
Escobar, M. R. and Edwards, P. R. (1968). *Can. J. Microbiol.*, **14**, 453–458.
Mollaret, H. and Nicolle, P. (1965). *C. r. hebd. Séanc. Acad. Sci. Paris*, **260**, 1027–1029.
Nicolle, P. (1973). *In* "Lysotypie und andere spezielle epidemiologische Laboratoriumsmethoden" (H. Rische, Ed.), pp. 377–387. VEB Gustav Fischer Verlag, Jena.

Nicolle, P., Mollaret, H. H. and Brault, J. (1968). *Symp. Series Immunobiol. Stand.*, **9,** 357–360.

Nicolle, P., Hamon, Y. and Diverneau, G. (1962). *Arch. Roum. Path. exp. Microbiol.*, **21,** 315–336.

Nicolle, P., Mollaret, H., Hamon, Y. and Vieu, J. F. (1967). *Ann. Inst. Pasteur (Paris)*, **112,** 86–92.

Nicolle, P., Mollaret, H. and Brault, J. (1969). *Arch. Roum. Path. exp. Microbiol.*, **28,** 1019–1027.

Nicolle, P., Mollaret, H. and Brault, J. (1973). *Microbiol. Immunol.*, **2,** 54–58.

Niléhn, B. (1969). *Acta path. microbiol. scand. Suppl. 206,* 1969.

Niléhn, B. (1973). *Microbiol. Immunol.*, **2,** 59–67.

Niléhn, B. and Ericson, C. (1969). *Acta path. microbiol. scand.*, **75,** 177–187.

Robbins, P. W. and Uchida, T. (1962). *Biochem.* **1,** 323–335.

CHAPTER IV

# Yersinia enterocolitica (Synonyms: "Pasteurella X", Bacterium enterocoliticum for Serotype 0-8)

STEN WINBLAD

*Department of Clinical Bacteriology, University of Lund,*
*Malmö General Hospital, Malmö, Sweden*

## I. IDENTITY

*Yersinia enterocolitica* is a Gram-negative rod, morphologically similar to *Pasteurella*. It grows on agar, blood agar and on media such as endoagar, SS-agar, desoxycholate agar and LSU agar (Juhlin and Ericson, 1961) that are routinely used for the diagnosis of pathogenic enterobacteria. Growth is, however, slow and 48 h incubation are needed to achieve good colony size. A temperature of 22°C (room temperature) results in better growth than a temperature of 37°C and that is important for diagnostic purposes. *Y. enterocolitica* is urease positive and phenylalanine negative, oxidase negative, catalase positive, usually sucrose positive, cellobiose positive and Voges-Proskauer positive only at 22°C. Isolates show motility in semi-solid medium which is best observed at 22°C. When fermentation of lactose is positive it is first detected after 3–4 days of growth. Cultures survive and even have the ability to multiply at 4°C. DNA hybridisation (Moore and Brubaker, 1975) has shown that *Y. enterocolitica* including all different O-serogroups constitutes a valid species different from *Y. pseudotuberculosis*, *Y. pestis*, *Escherichia coli* and other enterobacteria, but with a closer relationship to *Y. pseudotuberculosis* and *Y. pestis* than to other Gram-negative species.

## II. GENERAL BIOCHEMICAL PROPERTIES

A series of biochemical reactions applicable to Gram-negative bacteria has been used by Niléhn (1967 and 1969), and by Wauters (1970) and the results can be summarised as follows:

*Reactions always negative:* Oxidase reaction, hydrogen sulphide production, arginine decarboxylation, lysine decarboxylation, gelatin liquefaction, phenylalanine deamination, malonate utilisation, no acid formation from glycogen, inositol, melezitose, inulin, adonitol, arabitol, dulcitol, and erythritol.

*Reactions always positive:* Catalase production, urease test (25 and 37°C), growth in potassium cyanide medium, acid formation from d-glucose, fructose, mannose, mannitol, l-arabinose, d-galactose, maltose, d-ribose, levulose, cellobiose and β-galactosidase activity.

*Reactions negative with some exceptions:* Citrate utilisation at 37–22°C, Voges-Proskauer's reaction at 37°C, and acid formation from lactose, d-melibiose and d-raffinose.

*Reactions positive with some exceptions:* Motility in semi-solid agar at 22–25°C, nitrate reduction, ornithine decarboxylation, acid formation from l-sorbose, maltose, sucrose, sorbitol, arbutin, and Voges-Proskauer's reaction at 22°C.

*Reactions variable:* Lecithinase production, indole production, aesculin hydrolysis, and acid formation from salicin, d-xylose, trehalose, l-rhamnose, dextrin, glycerol, inositol, and amygdalin. Some of these variable reactions are of value for dividing *Y. enterocolitica* into biotypes and subgroups.

Differentiation between *Y. enterocolitica* and *Y. pseudotuberculosis* is illustrated in Table I. Differentiation from *Proteus* is best shown by the

## TABLE I

**Biochemical reactions of *Y. enterocolitica* and
*Y. pseudotuberculosis***

|  | Yersinia enterocolitica | Yersinia pseudotuberculosis |
|---|---|---|
| Voges-Proskauer at 25°C | + | − |
| Sucrose | +[1] | − |
| Rhamnose | −[2] | + |
| Cellobiose | + | − |
| Melibiose | −[3] | + |
| Sorbose | + | − |
| Indole | − + +[4] | − |
| Aesculin hydrolysis | − − + | + |
| Salicin | − − + | −[5] |

[1] With few exceptions
[2] Variable. Positive in "*Y. enterocolitica*-like" strains
[3] Positive for serogroup 17
[4] Three variations
[5] Positive after seven days incubation

negative phenylalanine reaction and from *Pseudomonas* by the negative oxidase reaction.

## III. ANTIGENS

An O-antigen of lipopolysaccharide nature is well studied and provides the basis for O-serogroups (Wauters, 1970; Wauters *et al.*, 1971; Wauters *et al.*, 1972; Winblad, 1967; Winblad, 1968; Winblad, 1973). H-antigen (flagellar-antigen) has been studied by Wauters (1970). It is possible that K-antigen also exists among some strains of O-serogroup 3 but this needs further elucidation.

### A. Preparation of O-antigen

Strains should be cultivated on blood agar† (1·4% peptonised agar) at 22–25°C for 48 h and suspended in saline. After autoclaving at 120°C for 20 min the centrifuged bacteria are washed once in saline and can be used as an antigen in an agglutination reaction against sera or for immunising rabbits to make anti-O-serum. Cell-free heated supernatant is also able to immunise rabbits for anti-O-serum production.

† Beef-extract 0·5%, agar 1·4%, NaCl 0·3%, sodium phosphate 0·2%, to which is added an equal volume of defibrinated blood from horses, sheep, rabbits, or humans.

## B. Immunisation schedule

For immunising rabbits a heated and washed suspension is used with a dose of $10^5$ cells per ml, starting with 0·5 ml and increasing to 1·0 ml on the 7th, the 14th, and the 21st day, all administered intravenously. An agglutinin titre of about 1/5000 will be established by the 28th day. The agglutination reaction is carried out at 52°C and the results read after 18–22 h.

## C. Preparation of H-antigen

Cultures should be grown at 22–25°C on 3% agar for 18 h in Petri dishes and then suspended in peptone water. After 5 h at room temperature the suspension is placed on 1% agar in large Petri dishes and incubated for 18 h at room temperature. After harvesting at 4°C the culture shows great motility after this treatment and is then treated with 0·5% formalin. O-serogroups 2 and 3 have fewer flagella (Niléhn, 1969) and repeated cultivation may be necessary (Wauters, 1970). After formolisation and centrifugation the culture is suspended in saline for immunisation or for use as an agglutination antigen. The agglutination reaction for H-agglutinin takes place at 37°C and is read after 2 h.

## D. Preparation of OH-antigen and living antigen

These are made in essentially the same way. OH-antigen means a suspended culture, killed by 0·5% formalin or 5% phenol, washed and used for immunisation of rabbits or as an antigen in an agglutination reaction at a temperature at 37°C overnight. Living cultures can also be used for immunising rabbits and cause no ill effects. Rabbit antisera prepared by OH-antigen or living antigens also show high agglutinin titres in reactions against O-antigens.

## E. Absorption of "common antigens"

Immunisation of rabbits with an O-antigen is an artificial and experimental method, never reflecting infection *in vivo*. Rabbit sera after immunisation with O-antigens of *Y. enterocolitica* show agglutination not only against *Y. enterocolitica* but also against many other Gram-negative species such as *Salmonella* of groups A, B, C and D, *E. coli*, *Proteus* and *Pseudomonas aeruginosa*. If rabbits are immunised with O-antigens of *Salmonella* or *E. coli* the immune sera show agglutinins not only against the species used for immunisation, but also against *Y. enterocolitica* and the other Gram-negative species, listed above. This reflects the presence of a "common" antigen, known to be part of the lipopolysaccharide structure of many Gram-negative rods.

Absorption of antiserum against *Y. enterocolitica* O-antigens with *E. coli* 0–14 or *Salmonella* O antigens results, however, in a serum containing specific agglutinins for *Y. enterocolitica*. Absorption with a heterologous O-group of *Y. enterocolitica* results not only in the disappearance of the "common" agglutinins but also of the non-type specific agglutinins against *Y. enterocolitica*. Such absorption processes are recommended for making specific O-antisera of *Y. enterocolitica*.

Immunising with OH-antigen or living antigen never results in agglutinins against the "common" antigen and simulates better the occurrence of infections in man and animals. The detection of O-agglutination with human patients' sera is, for that reason, considered as a "specific" response to an infection. This is true when using ordinary tube agglutination methods but not when using passive haemagglutination with erythrocytes coated with O-antigens. In the latter test antibodies against the "common" antigen will be observed (Macland and Digranes, 1975a; and b).

To prepare specific O-antisera by absorption O-serogroup strains are cultured on blood agar at room temperature for 48 h, harvested, suspended in saline, autoclaved at 120°C for 20 min, washed in saline and centrifuged. The pellet is mixed with undiluted rabbit immune serum that is to be absorbed, held for 1 h at 37°C and re-centrifuged. The absorbed serum must be tested to ensure that it has been freed from agglutinins against the absorbing strain. The process may be repeated if necessary. The serum is then tested to determine the dilution that gives good rapid slide agglutination against the specific group strain. Living cultures can be used in such slide agglutinations. The original culture must also be checked to ascertain that it has not become transformed into the R-form. Table II lists strains, which are recommended for making such group specific O-antisera.

## F. Cross antigenicity

This is seen in some serogroups of *Y. enterocolitica* (see below). Cross antigenicity against *Y. pseudotuberculosis*, *Y. pestis* or *Francisella tularensis* is not observed. Between *Y. enterocolitica* serogroup 9 and *Brucella* there exists a very marked cross antigenicity. This phenomenon has been intensively studied (Ahvonen *et al.*, 1969; Corbel and Cullen, 1970; Corbel, 1973; Hurvell, 1973). Absorption for separating specific antibodies is only successful with rabbit hyperimmune sera but unfortunately not in sera from humans or animals infected with *Y. enterocolitica* or with *Brucella*. This complicates the serological diagnosis of such infections.

## TABLE II

Strains representing different O-groups of *Y. enterocolitica*, useful for immunisation and strains for absorption for making specific O-antiserum

| O-group | Antigens | No Inst. Past.† | Name | Source | Absorbing strain | Anti-O-sera |
|---|---|---|---|---|---|---|
| 1 | I, II, III | 64 | Becht 57/6 | Chinchilla | 14 | 1 |
| 2 | II, III | 14 | Lucas 404‡ | Hare | 134 | (1), 2 |
| 3 | III | 134 | Winblad My O | Human | 96 | (1, 2) 3 |
| 4 | IV | 96 | Knox | Chinchilla | 64 | 4 |
| 5a | Va | 1402 | Borges | Cow milk | 97 | 5a |
| 5b | Vb | 97 | Knox | Chinchilla | 1402 | 5b |
| 6 | VI | 102 | Bojsen-Möller | Human | 64 and 134 | 6 |
| 7 | VII, VIII | 107 | Borg Pedersen | Guinea pig | 102 and 106 | 7 |
| 8 | VIII | 106 | Schleifenstein | Human | 134 | (7), 8 |
| 9 | IX | 336 | Niléhn My 79 | Human | 134 and 123 | 9 |
| 10 | X | 474 | Wauters | Human | 96 and 614 | 10 |
| 11 | XI | 105 | Kristensen | Human | 64 and 103 | 11 |
| 12 | XII | 490 | Lucas 63 | Hare | 474 | 12 |
| 13 | XIII | 553 | Wauters | Human | 107 and 480 | 13 |
| 14 | XIV | 480 | Graux | Human | 490 | 14 |
| 15 | XV | 614 | Esseveld 188/69 | Human | 64 | 15 |
| 16 | XVI | 867 | Graux | Human | 96 | 16 |
| 17 | XVII | 955 | Lassen 333B | Water | 134 | 17 |
| 18 | XVIII | 846 | Weaver 4542 | Human | 474 | 18 |
| 19 | XIX | Fy 50 | Ahvonen 534/71 | Human | 96 | 19 |
| 20 | XX | 845 | Weaver 4403 | Human | 64 and 474 | 20 |

† Number from the collection of *Y. enterocolitica* strains of Institut Pasteur, with exception of group 19.

‡ Immunising with living culture.

## G. H-antigens

Wauters *et al.* (1971, 1972) have described labile antigens that are assumed to be situated on the flagellae of *Y. enterocolitica*. In all nineteen H-factors named a–i and k–t have been described. The majority of the O-groups have one H-antigen each. This is specific for that group and is not shared by other groups. Exceptions are the O-groups 1, 2, 3, 5, 6, 7, 8, and 10 which may be further subdivided on the basis of their H-antigens.

## H. K-antigens

In group 0–10 there are strains which possess a K-antigen (Wauters, 1970) making the strains inagglutinable in O-antisera, unless treated first at 100°C for 30 min.

## IV. DIFFERENT TYPES OF *Y. ENTEROCOLITICA*

Within the species *Y. enterocolitica* different O-serogroups exist as shown by absorption. Another subdivision into *biotypes* is based on variations in biochemical reactions, and *phage types* have also been defined. H-antigens have been studied and compared with O-serogroups (Wauters, 1970).

### A. O-serogroups

The O-serogrouping scheme is based on the studies of Winblad (1967, 1968 and 1973) and of Wauters and his collaborators (1970, 1971 and 1972) and consists at present of 26 O-group antigens (numbers 1–26). Many strains, especially aquatic strains, are not included in this scheme and are as yet untypable. Cross antigenicity between serogroups 1, 2, and 3, and between 7 and 8 exist as is shown in Table III. Recent studies have shown a complex antigenic structure in some strains, which may lead to a further subdivision of the original O-groups into subgroups and provide better information concerning cross reactions between present O-groups. O-serogroup 1 includes practically all strains cultivated from chinchillas. O-serogroup 2 strains are cultivated from hares and goats. These strains are slow growing, have a tendency to transform into R-forms, and show poor fermentation ability. O-serogroup 3 includes the ordinary human pathogenic strains from Europe, Africa, South America and Asia. O-serogroup 8 includes the dominant strains of *Y. enterocolitica* from human cases in USA and has not been observed in Europe. O-serogroup 9 is pathogenic for humans and has been observed in several countries of Europe, especially in Finland (Ahvonen, 1972). Division based on H-antigens leads to the definition of the same groups as does O-antigen analysis.

### B. Biotypes

Niléhn (1967 and 1969) and Wauters (1970) subdivide *Y. enterocolitica* into five biotypes, as illustrated in Table IV.

### C. Phage types

These have been studied by Mollaret and Nicolle (1965) and Nicolle *et al.* (1967, 1972) and by Niléhn *et al.* (1969) and are based on lysogenicity of some strains and on lysotyping. This system has not yet achieved the same epidemiological importance as O-serogrouping. For methods used in lysotyping the reader is referred to the original work by Nicolle *et al.* (1967, 1952) and to this Volume Chapter II.

## TABLE III

O-agglutinin-titre of rabbit immune sera against *Y. enterocolitica* before and after cross absorption

| Antisera | Bacterial O-group-antigens | | | | | | | | | |
|---|---|---|---|---|---|---|---|---|---|---|
| | 1 | 2 | 3 | 4 | 5a | 5b | 6 | 7 | 8 | 9 |
| Homologous | 5120 | 5120 | 2560 | 5120 | 2560 | 5120 | 2560 | 5120 | 5120 | 5120 |
| Absorbed with group 1 | 0 | 0 | 0 | 1280 | 2560 | 5120 | 2560 | 2560 | 2560 | 2560 |
| Absorbed with group 2 | 80 | 0 | 0 | 2560 | 2560 | 2560 | 2560 | 2560 | 2560 | 2560 |
| Absorbed with group 3 | 320 | 2560 | 0 | 2560 | 2560 | 5120 | 2560 | 2560 | 2560 | 2560 |
| Absorbed with group 4 | 640 | 2560 | 2560 | 0 | 2560 | 5120 | 2560 | 2560 | 2560 | 2560 |
| Absorbed with group 5a | 640 | 2560 | 2560 | 1280 | 0 | 1280 | 2560 | 2560 | 2560 | 2560 |
| Absorbed with group 5b | 640 | 2560 | 2560 | 1280 | 1280 | 0 | 2560 | 2560 | 2560 | 2560 |
| Absorbed with group 6 | 1280 | 5120 | 2560 | 2560 | 2560 | 2560 | 0 | 2560 | 1280 | 1280 |
| Absorbed with group 7 | 640 | 5120 | 2560 | 1280 | 2560 | 2560 | 2560 | 0 | 0 | 2560 |
| Absorbed with group 8 | 640 | 5120 | 2560 | 1280 | 2560 | 2560 | 2560 | 160 | 0 | 2560 |
| Absorbed with group 9 | 320 | 5120 | 2560 | 2560 | 2560 | 2560 | 2560 | 640 | 2560 | 0 |

TABLE IV

**Biotypes modified after Niléhn (1969)**

|  | Biotypes | | | | |
|---|---|---|---|---|---|
|  | 1 | 2 | 3 | 4 | 5 |
| Salicin | + | − | − | − | − |
| Aesculin | + | − | − | − | − |
| Lecithinase | + | − | − | − | − |
| Indole | + | + | − | − | − |
| d-Xylose (48 h) | + | + | + | − | − |
| Lactose O/F (48 h) | + | + | + | − | − |
| Nitrate | + | + | + | + | − |
| Trehalose | + | + | + | + | − |
| Sorbitol | + | + | + | + | − |
| Ornithine | + | + | + | + | − |
| β-galactosidase | + | + | + | + | − |

## D. The ecological significance of O-serogroups and biotypes

The ecological significance of the different groups as studied by the author is summarised in Table V. The findings are based on an investigation of about 900 strains from different sources and from different geographical districts. The results suggest a logical separation into subgroups. This subgrouping is rather similar to that of Knapp and Thal (1972), based on only a few strains, but differs in some details. Some biochemical reactions, which correlate closely with O-antigen groupings, provide a useful basis for characterisation.

The indole reaction is particularly useful. Fermentation of d-xylose is negative for strains of O-serogroup 3, the most common human pathogenic strains, and for the strains from hares and goats belonging to O-serogroup 2. The last mentioned strains are also the only trehalose negative strains (Subgroups I and II). All these strains never have lecithinase. Lecithinase is constantly combined with indole positivity with the exception of O-serogroup 9, which shows indole positivity only after 3–4 days of growth. In subgroup II are collected O-serogroup 1 (chinchilla strains), O-serogroups 11 and 12 and O-serogroup 9, all d-xylose positive and lecithinase negative. Subgroups IV, V, and VI include all indole-positive strains.

As a separate subgroup (IV) O-serogroup 8 is to be noted. All of these strains are cultivated from human cases in USA. The strains are indole positive and do not ferment salicin or aesculin. They are biochemically as well as serologically separate from other *Y. enterocolitica* strains and

## TABLE V

### Relationship between biochemistry, serogroups, subgroups and sources

| Subgroup | Indole | Aesculin | Salicin | Voges-Proskauer 22°C | Rhamnose | Sucrose | d-Xylose | Trehalose | Melibiose | Lecithinase | Biotype | O-Serogroup | Sources | No. of strains studied |
|---|---|---|---|---|---|---|---|---|---|---|---|---|---|---|
| *Yersinia enterocolitica* | | | | | | | | | | | | | | |
| I. (d-Xylose neg.) | − | − | − | + | − | + | − | + | − | − | 4 | 3 | Humans, pigs, monkeys | 564 |
| II. (d-Xylose neg., Trehalose neg.) | − | − | − | − | − | + | − | − | + | − | 5 | 2 | Hares, goats | 51 |
| III. (d-Xylose pos.) A | − | − | − | + | − | + | + | + | − | − | 3 | 1, 4a, 5b | Chinchilla | 62 |
| B | − | − | − | + | − | − | + | + | − | − | 3 | 11, 12 | Animals, humans | 10 |
| C | (+) | − | − | + | − | + | + | + | − | − | 3 | 9 | Humans | 77 |
| IV. (Indole pos.) | + | − | − | + | − | + | + | + | − | + | 2 | 8 (18, 20, 22) | Humans in U.S.A. | 42 |
| *Yersinia enterocolitica-like strains* | | | | | | | | | | | | | | |
| V. (Indole, Aesculin, Salicin pos.) | + | + | + | + | − | + | + | + | − | + | 1 | 4b, 5a, 6, 7, 7/13, 10, 13, 14, 15, 19 | Animals, Water | 74 |
| VI. A (Rhamnose pos.) | + | + | + | + | + | + | + | + | − | + | 1 | 16, 23–26 | Water | 22 |
| B (Melibiose pos.) | + | + | + | + | + | + | + | + | + | + | 1 | 17 | Humans in U.S.A. | 15 |

correspond to the early *"Bacterium enterocoliticum"* of Schleifenstein and Coleman (1939a and b).

Strains in subgroups V and VI are cultivated from humans, animals and water. Their pathogenicity for humans is uncertain. Subgroup V strains do not ferment rhamnose. In subgroup VI are collected all the rhamnose positive strains. O-serogroup 17 ferments melibiose and resembles *Y. pseudotuberculosis.* Human infections are, however, observed with O-serogroup 17 (Bottone *et al.,* 1974).

## V. DIAGNOSTIC METHODS

*Y. enterocolitica* is routinely isolated from stools or from extirpated appendices. A few cases have been described in which *Y. enterocolitica* was isolated from blood or urine. Routine cultivation may be carried out using ordinary media suitable for pathogenic enterobacteria. SS-agar seems to be suitable and better than endoagar. Cultivation demands incubation for 48 h and is best at room temperature. A good selective method includes primary cultivation of the material in ordinary peptone broth at 4°C for 3–5 days (Wauters, 1970). *Y. enterocolitica* grows well at this temperature in contrast to *E. coli* and other enterobacteria. Wauters recommended selenite broth after Leifson (1936) supplemented by 0·007% Malachite green as a selective method (Wauters, 1970). Suspect colonies which are urease positive, catalase negative, motile in semi-solid agar only at room temperature, and have a positive Voges-Proskauer reaction only after incubation at room temperature indicate the presence of *Y. enterocolitica.* In fermentation tests they should be sucrose positive and cellobiose positive. The definitive diagnosis must be confirmed by slide agglutination of the living culture with absorbed specific O-antisera (commonly anti-3, anti-9 or in USA anti-8). If *Y. enterocolitica* is suspected but agglutination with the common anti-O-sera is negative, a more intensive study of fermentation and slide agglutination against other specific anti-O-group sera should be performed.

Development of antibodies in a patient's serum is the basis for another very important diagnostic method in cases of human infections. Agglutinins detected by the ordinary Widal technique against suspended, heat treated and washed bacteria of relevant serotypes are present after an infection and such agglutinin reactions seem to be very specific when using tube agglutination techniques, but not when using passive haemagglutination techniques.

## VI. PATHOGENICITY

O-serogroup 1 is cultivated from sick chinchillas, O-serogroup 2 from dead hares (Mollaret and Lucas, 1965) and goats. O-serogroups 3, 8, and

9 seem to be human pathogens. Experimental studies have shown that O-serogroups 3 and 9 have no pathogenicity for ordinary laboratory animals such as rabbits, guinea-pigs, rats or mice (Mollaret and Guillon, 1965). O-serogroup 8 may, according to Carter *et al.* (1974) show pathogenicity for mice. Human pathogenic strains have also been cultivated from pigs, dogs and monkeys (Mollaret *et al.*, 1970). Investigations of the faecal flora from *Microtus*, made in Norway and in France, have resulted in cultivation of *Y. enterocolitica*, but mostly, however, of human non-pathogenic O-serogroups and very rarely of O-serogroup 3 (Alonso and Bercovier, 1975; Knapperud, 1975). Whether O-serogroups other than 3, 8, and 9 are pathogenic for man is uncertain. Development of antibodies against such O-serogroups as 4, 5a, 5b, 6, and 7 in human sera has not been observed in spite of the cultivation of such strains from stools. The "*Y. enterocolitica*-like" strains seem to be non-pathogenic for man and most likely belong to the natural flora of water. The clinical panorama for human infections with O-serogroups 3, 8, and 9 is a rather broad one (Ahvonen, 1972; Mollaret, 1966; Niléhn, 1969; Winblad *et al.*, 1966; Winblad, 1973). Small children suffer from gastro-enteritis, older children show acute terminal ileitis, mesenterial lymphadenitis and pseudo-appendicitis, all with clinical symptoms mimicking appendicitis. Enteritis is the most general symptom. Symptoms such as erythema nodosum and arthritis are not infrequent complications. Even cases showing septicaemia, nephritis, conjunctivitis or uveitis have been described. But in general infection with this microbe does not cause a very serious disease.

## VII. ANTIBIOTIC SENSITIVITY AND THERAPY

*Y. enterocolitica* is resistant to Penicillin, Methicillin, Ampicillin, Oleandomycin, Novobiocin and Fucidin (Niléhn, 1969). Resistance to the Penicillin group depends on the ability of the strains to produce beta-lactamase (Cornelius, 1975). All strains are sensitive to Streptomycin, Tetracyclin, Oxytetracycline, Chloramphenicol, Nitrofuradantoin, sulphonamides, Gentamicin and Nalidixic acid. Sensitivity to Neomycin and Kanamycin is variable. Resistance factors can sometimes be transferred to this species but this is relatively unusual (Cornelius, 1975). If for clinical reasons antibiotic therapy is necessary Tetracyclin may be used. Strains of O-serogroup 8 are, however, sensitive to Penicillin, Methicillan and Ampicillin.

## VIII. EPIDEMIOLOGY

Human infection is usually food borne. Family cases and clustered cases in institutions such as boys' schools have been described (Asokawa *et al.*, 1973; Gutman *et al.*, 1973). *Y. enterocolitica* of serogroup 3 has been isolated from pork in several countries and it is possible that human infections can come from pork (Zen-Yoji *et al.*, 1974). How pigs become infected is not yet known.

## IX. GEOGRAPHICAL FACTORS

Human cases of serogroup 3 are well known all over Europe. In Scandinavia, the Netherlands, Belgium, France, Hungary and Czechoslovakia such infections have been particularly well studied. No cases, however, have been reported from the British Isles. Monkeys in Brazil have been reported as infected by O-serogroup 3 (Mollaret *et al.*, 1970). Many cases are reported from Japan and a few cases from USSR, South Africa, Australia and the Congo (Makulu *et al.*, 1969). In Canada (Toma and Lafleur, 1974) the disease is well known with infections not only caused by O-serogroup 3. It is remarkable that serogroup 8 is reported only from USA and that no cases involving this serogroup have been observed in Europe.

Yersiniosis enterocolitica must be regarded as a fairly common infectious disease. The main problem at the present time is to find the vectors of the infections and to identify possible carriers in the animal world.

## REFERENCES

Ahvonen, P., Jansson, E., and Aho, K. (1969). *Acta Path. microbiol. scand.*, **75,** 291–295.

Ahvonen, P. (1972). *Ann. clin. Res.*, **4,** 30–48.

Alonso, J. M., and Bercovier, H. (1975). *Med. Mal. Infect.*, **3,** 180–181.

Asokawa, Y., Akahani, S., Kagata, N., Nogichi, M., Sakazaki, R., and Tamura, K. (1973). *J. Hyg.*, **71,** 715–723.

Bottone, E. J., Chester, B., Malowany, M. S., and Allerhand, J. (1974). *Appl. Microbiol.*, **27,** 858–861.

Carter, Ph.B., and Collins, F. M. (1974). *Infect. Immunity*, **9,** 851–857.

Corbel, M. J., and Cullen, G. A. (1970). *J. Hyg.*, **68,** 519–530.

Corbel, M. J. (1973). *J. Hyg.*, **71,** 309–323.

Cornelius, G. (1975). *J. gen. Microbiol.* (in press).

Gutman, L. F., Ottesen, E. N., Quan, T. J., Noce, O. S., and Katz, S. L. (1973). *New Engl. J. Med.*, **288,** 1372–1377.

Hurvell, B. (1973). *Acta path. microbiol. scand.*, *B*, **81,** 105–112.

Juhlin, I., and Ericson, C. (1961). *Acta path. microbiol. scand.*, **52,** 185–200.

Knapperud, G. (1975). *Acta path. microbiol. scand.*, *B*, **83,** 335–342.

Knapp, W., and Thal, E. (1972). *Zbl. Bakt. Hyg. I Abt. Orig.*, *A*, **223,** 88–105.

Leifson, E. (1936). *Am. J. Hyg.*, **24,** 423–432.

Macland, J., and Digranes, A. (1975a). *Acta path. microbiol. scand.*, *B*, **83,** 382–386.

Macland, J., and Digranes, A. (1975b). *Acta path. microbiol. scand.*, *B*, **83,** 451–456.

Mollaret, H. H., and Guillon, J. C. (1965). *Ann. Inst. Pasteur*, **109**, 608–613.

Mollaret, H. H., and Lucas, A. (1965). *Ann. Inst. Pasteur*, **108**, 121–216.

Mollaret, H. H. (1966). *Path. Biol.*, **14**, 981–990.

Mollaret, H. H., Giorgi, W., Matera, A., Portana de Castro, A., and Guillon, J. C. (1970). *Res. Med. Vet.*, **146**, 919–924.

Mollaret, H. H., and Nicolle, P. (1965). *C. r. hebd. Séanc. Acad. Sci., Paris*, **260**, 27–1029.

Moore, R. L., and Brubaker, R. R. (1975). *Int. J. System. bact.* (in press).

Nicolle, P., Mollaret, H. H., Hamon, Y., and Vieu, J. F. (1967). *Ann. Inst. Pasteur*, **112**, 86–92.

Nicolle, P., Mollaret, H. H., and Brault, J. (1972). *Bull. de l'Academ. de Med., Paris*, **156**, 712–721.

Niléhn, B. (1967). *Acta path. microbiol. scand.*, **69**, 83–91.

Niléhn, B. (1969). *Acta path. microbiol. scand. suppl.*, **206**.

Niléhn, B., and Ericson, C. (1969). *Acta path. microbiol. scand.*, **75**, 177–187.

Niléhn, B. (1969). *Acta path. microbiol. scand.*, **77**, 527–537.

Schleifstein, J., and Coleman, M. (1939). *N. Y. St. J. Med.*, **39**, 1749–1753.

Schleifstein, J., and Coleman, M. (1943). *Ann. Rep. Div. Lab. Res. N. J. St. Dept. Hlth*, **56**.

Toma, S., and Lafleur, L. (1974). *Appl. microbiol.*, **28**, 469–473.

Wauters, G. (1970). "Contribution à l'étude de *Yersinia enterocolitica*". Vander, Louvain.

Wauters, G., Le Minor, L., and Chalon, A. M. (1971). *Ann. Inst. Pasteur*, **120**, 631–642.

Wauters, G., Le Minor, L., Chalon, A. M., and Lassen, J. (1972). *Ann. Inst. Pasteur*, **122**, 951–956.

Winblad, S., Niléhn, B., and Sternby, N. H. (1966). *Brit. Med. J.*, **II**, 1363–1366.

Winblad, S. (1967). *Acta path. microbiol. scand. suppl.*, **187**, 115.

Winblad, S. (1968). *Int. Symp. Pseudotuberculosis, Symp. Ser. Immunbiol. Standard*, **9**, 337–342. Karger, Basel/New York.

Winblad, S. (1973). *Contr. Microbiol. Immunol.*, **2**, 27–29, Karger, Basel/New York.

Winblad, S. (1973). *Contr. Microbiol. Immunol.*, **2**, 129–131, Karger, Basel/New York.

Zen-Yoji, H., Sakai, S., Maruyama, T., and Yanagawa, Y. (1974). *Jap. J. Microbiol.*, **18**, 103–105.

CHAPTER V

# Principles and Practice of Typing
# *Vibrio cholerae*

S. MUKERJEE

*WHO International Reference Centre for Vibrio Phage Typing, Calcutta*

# I. GENERAL INTRODUCTION

Cholera is a scourge of human beings. A specific gastro-intestinal infective disease, it is known to have involved most countries of the world in five pandemics originating from its Asiatic home in India. According to the estimate of Pollitzer (1959), over one hundred million lives were lost in these pandemics.

The cholera pandemics originating in Asia were known as Asiatic cholera. The causative organism of this disease was known as classical *Vibrio cholerae*. Another type of this organism was isolated from the intestine of six Mecca pilgrims in the El Tor quarantine station (Gotschlich, 1906). These vibrio strains closely resembled the classical type of *V. cholerae* except that they were able to haemolyse sheep or goat erythrocytes, which property was missing in the classical strains. These, commonly known as El Tor vibrios, have been recently designated as *Vibrio cholerae* biotype *eltor* by the IAMS Subcommittee on Taxonomy of Vibrios Feeley 1969. Subsequently the El Tor vibrios were isolated from mild cases of diarrhoea, healthy individuals, and water sources in the Middle East countries and India in the absence of cholera. El Tor vibrios therefore came to be considered non-pathogenic.

In September 1937, a cholera-like epidemic was observed in South-West Celebes in Indonesia. The disease was severe in nature and was found to be caused by El Tor vibrios. This disease was named choleriformis El Tor by Van Loghem (1926) and paracholera El Tor by de Moor (1949) and is now better known as cholera El Tor. From 1937 to 1958, the disease remained almost exclusively localised in South Celebes. By the second half of 1961 extensive territories in the Far East including Hong Kong, Macao, Sarawak and the Philippines were involved in outbreaks of a cholera-like disease. Mukerjee and Guha Roy (1962) were the first to report isolates of *V. cholerae* from outbreaks in Hong Kong, Macao and Manila as belonging to the *eltor* biotype, a finding subsequently confirmed by Felsenfeld (1963a, b); Lee Hin (1963, pers. comm.) and Acurin (1963, pers. comm.). Within a decade, cholera El Tor spread over the West Pacific Islands and South, South East and Central Asian countries in a pandemic form and invaded most of the countries in Africa and also spread to parts of Europe. This seventh pandemic of cholera caused by *V. cholerae* biotype *eltor* is now raging unabated in three continents. At present, there is little hope that an early containment of the pandemic will be possible by application of known hygienic measures and the international laws on cholera control. Countries which are now free from cholera may get involved at any time. It has thus emerged as one of the most important disease problems for the welfare of mankind.

Cholera is a disease of bacterial origin. Much important research followed Robert Koch's discovery in 1884 of a comma-like vibrio in the stools of cholera patients. It is not only important to recognise the disease clinically, but also to attempt as complete an identification of the causative organism as possible using various epidemiological typing procedures. Routine identification and typing of *V. cholerae* in diagnostic laboratories at national levels in co-operation with the WHO Collaborative Centre for Research and Reference on Vibrios, established at Calcutta, will facilitate measures aimed at control of the disease. This Chapter provides a detailed account of practical procedures currently available for the purpose of characterisation, as well as a working background of the relevant theoretical knowledge.

## II. DEFINITION OF *VIBRIO CHOLERAE*

*V. cholerae* is a motile Gram-negative, asporogenous, single curved or rigid rod with a single polar flagellum. It is indophenol oxidase positive and produces acid without gas from glucose.

*V. chlorae* usually does not ferment lactose within 24 hours. Regarding saccharolytic activity, *V. cholerae* (O group I) belongs to Heiberg group 1 (Heiberg, 1934) fermenting glucose, saccharose, and mannose, but not arabinose. According to recent recommendations of the Taxonomic Subcommittee of the International Association of Microbiological Societies, a strain can be called *V. cholerae* only if it produces lysine and ornithine decarboxylase, but not arginine dehydrolase or hydrogen sulphide (detectable in TSI or Kligler's iron-agar medium).

Until recently, only the O-group I vibrios were considered true cholera vibrios. The non-agglutinable vibrios were known as NAG vibrios or NCV (Non-cholera vibrios). Recently, however, there is evidence to indicate that NAG vibrios may be as choleragenic as their agglutinable counterparts. They also fulfil the minimum requirements for classification as *V. cholerae*. The second meeting of the IAMS Subcommittee on Taxonomy of vibrios has recommended that *V. cholerae* should also include the NAG vibrios.

The O-group I vibrios have been further classified into two biotypes, classical and *eltor*, based on differences demonstrable by a number of tests. The biotypes are usually stable in the laboratory and are of epidemiological significance.

## III. BIOTYPES OF O GROUP I *V. CHOLERAE*

### A. Historical

As early as 1906, Kraus and Pribram found that the El Tor vibrios first isolated by Gotschlich (1906) in the El Tor quarantine station in the Sinai peninsula differed from the classical type of *V. cholerae* in being able to haemolyse sheep or goat erythrocytes. The El Tor vibrios were different from classical cholera vibrios in a number of characteristics. The haemolysis test, which for a long time was used as the only criterion for distinguishing the two types of cholera vibrios, was later found inadequate for the purpose. El Tor strains were reported which were non-haemolytic at the time of isolation, but became haemolytic after a few subcultures (de Moor, 1949; Tanamal, 1959; Mukerjee and Guha Roy, 1961a). There had also been instances when El Tor strains were found to be non-haemolytic ever since isolation (de Moor, 1963). In view of this variation in the haemolytic property of El Tor vibrios, a number of tests have been developed with a view to differentiating the biotypes of O group I cholera vibrio. The more important of these include chicken erythrocyte agglutination (Finkelstein and Mukerjee, 1963), and sensitivity to Polymyxin (Gan and Tjia, 1963) and to group IV phage (Mukerjee, 1963b). Biotypes are stable in the laboratory. Since the emergence of cholera El Tor in epidemic and pandemic forms, biotyping has assumed great epidemiological importance.

## B. Action of group IV bacteriophages on the *eltor* biotype of *V. cholerae*

Studies in our laboratories have shown that a group IV *V. cholerae* typing phage, 149, at its routine test dose (RTD) is lytic for all strains of the classical biotype of *V. cholerae*, but all *eltor* biotype strains are resistant to it. Based on this observation, a test has been developed using this phage for differentiating the two biotypes of O group I vibrios (Mukerjee, 1960, 1963b; Mukerjee and Guha Roy, 1961a, b). The value of this test has been confirmed by other workers including Barua and Gomez (1967). Monsur *et al.* (1965) studied the effect of group IV phage on El Tor vibrios and concluded that "Although group IV phage attacks El Tor vibrios, no indication of phage multiplication during this process has been found. The results do not detract from the value of phage IV as a taxonomic tool because the behaviour of phage towards true cholera and El Tor vibrios is different. Because the reported insusceptibility of El Tor strains is only relative, the concentration of phage used in the test can be critical. Such resistance and susceptibility should be expressed in terms of plaque-forming units of bacteriophages".

TABLE I

**Characteristics of the two cholera vibrio biotypes**

| Test | Haemolytic for sheep erythrocytes | Chicken erythrocytes agglutination | Polymyxin B sensitivity | Group IV phage sensitivity |
|---|---|---|---|---|
| Biotypes | | | | |
| 1. classical | — | — | + | + |
| 2. *eltor* | + | + | — | — |

Since the biotyping scheme is closely related to the phage typing scheme, the merits of these tests for identification of the two biotypes of cholera vibrios have been discussed separately, as a result of studies on a large series of strains.

## C. Biotyping methods for *V. cholerae* O group I

### 1. *Haemolysis test*

The original method of carrying out the haemolysis test described by

Greig (1914) has undergone various modifications and no standard technique was generally recognised until 1963 when Feeley and Pittman investigated the factors involved. They recommended the use of over-night cultures at 37°C in heart infusion broth (HIB), pH 7·1 to 7·4, mixed with equal volumes of a 1% suspension of sheep erythrocytes in normal saline and incubated at 37°C for 2 hours. Readings are taken after holding the tubes in the refrigerator overnight. Consistent results can be obtained by this test even with weakly haemolytic strains. However the authors found that haemolysin, which is produced by El Tor vibrios, is unstable when there is an alkaline shift in the pH of the media. This can be prevented by adding 1% glycerol to the heart infusion broth (GHIB) (Barua and Mukherjee, 1964). This modification has proved useful in the identification of El Tor strains which was difficult by the original method (Sanyal et al., 1972) and is recommended for routine testing.

## 2. Haemagglutination test

This test has been most extensively used in recent years as a simple, rapid and dependable test for differentiating the eltor and classical biotypes of V. cholerae. Most workers have used chicken erythrocytes, but sheep, goat and human erythrocytes have been used successfully (Barua and Mukerjee, 1964; Zinnaka et al., 1964). The blood may be collected in Alsever's solution or heparinised or defibrinated (Finkelstein and Muker-jee, 1963). Broth cultures of the test strains are reportedly unsatisfactory for haemagglutination (Barua and Mukherjee, 1964). The tests described here are for suspensions of overnight nutrient agar cultures (37°C) in normal saline tested with chicken erythrocytes.

(a) *Rapid slide test method:* (Finkelstein and Mukerjee, 1963). A microscope slide is marked across with approximately 20 parallel lines drawn with a wax based marking pen. Saline is spotted into the channels with a 3 mm diameter loop. Fresh test culture is emulsified to give a thick suspension. A loopful of 2·5% chicken red blood cells in normal saline is then added and mixed by tilting the slide back and forth. Clumping of erythrocytes occurs almost immediately, and is observed during one minute using transmitted light. The use of a stereoscopic microscope is particularly useful for confirmation of weak or doubtful reactions. Degraded or rough strains agglutinate erythrocytes satisfactorily.

(b) *Tube method:* To 0·5 ml volume of test culture dilutions, are added 0·5 ml of a 1% saline suspension of red blood cells. The tubes are shaken and allowed to stand at room temperature for 1–2 hours. The red cells settle at the bottom of the tubes in a characteristic fashion. A positive reaction is indicated by a reticular precipitation of the red cells, in contrast to the dense central button formed in negative cases.

### 3. *Polymyxin B sensitivity test*

Gan and Tjia (1963) reported the difference in sensitivity of the classical and *eltor* biotypes of *V. cholerae* strains to Polymyxin B at 50 μg/disc. These discs are placed on a lawn of the test bacterium prepared with its overnight broth culture on nutrient agar. The plates are incubated at 37°C for 24 hours. Sensitive strains show zones of inhibition around the discs.

Roy *et al.* (1965) obtained better and more clear cut results by spotting 2 hour broth cultures of the test strains on nutrient agar containing 15 μg/ml Polymyxin B. Only the resistant *eltor* biotype grows at this concentration.

Barua and Gomez (1967) recommended that the pH of the nutrient agar plates should be adjusted to between 7·0 and 7·6. Known sensitive and resistant strains should be included in each test.

### 4. *Group IV phage sensitivity test*

Mukerjee (1963b) reported all the classical biotype *V. cholerae* strains to be sensitive to the group IV cholera phage, while none of the biotype *eltor V. cholerae* strains is lysable by this phage at its routine test dilution (RTD) determined on the propagating strain Ogawa 154. A test was thus developed for differentiating the two biotypes of *V. cholerae* by using a large series of strains belonging to each biotype.

One loopful of a four hour nutrient broth culture (37°C) of the test strain is smeared on nutrient agar to a diameter of approximately 1·25 cm. One 3 mm loopful of the phage at its RTD is spotted on each of the smears. The plate is then incubated at 37°C for 24 hours. Classical biotype strains will show clear evidence of lysis, while growth of the *eltor* biotype is confluent.

## IV. THE COMPARATIVE VALUES OF PHAGE IV SENSITIVITY WITH OTHER KNOWN TESTS FOR DIFFERENTIATING THE TWO O GROUP I VIBRIOS

Gotschlich (1906) isolated vibrio strains agglutinable with cholera O serum from the intestine of 6 Mecca pilgrims. The patients came from an area which was completely free from cholera. This caused confusion regarding bacterial identity as true cholera vibrios. However, Kraus and Prantschoff (1906) found the strains to differ from *V. cholerae* by their production of a soluble haemolysin and a lethal exotoxin. Since then haemolysis testing has been utilised to differentiate the two types of agglutinable vibrios. The El Tor vibrios are usually haemolytic for sheep or goat erythrocytes in the Greig haemolysis test in which the classical strains are non-haemolytic.

The El Tor vibrios closely resemble classical *V. cholerae*. Both organisms exhibit the general characteristics of vibrios (Breed *et al.*, 1957), the biochemical reactions of Heiberg Group 1 (Heiberg, 1934) and agglutinability by cholera O serum (Gardner and Venkatraman, 1935). The difference in haemolytic activity of the two types is generally considered as a differentiating criterion.

It has been reported by de Moor (1949) that the haemolytic power of El Tor vibrios is retarded in some newly isolated strains and therefore may be overlooked. In fact, El Tor strains may give alternatively positive and negative results. El Tor vibrio strains may be non-haemolytic at the time of isolation, but haemolytic after a few subcultures (de Moor, 1949; Tanamal, 1959; Mukerjee and Guha Roy, 1961a). There have also been instances where a strain of El Tor vibrio was found to be stably non-haemolytic (Mukerjee, unpublished results). Strains Nos. Bandung 2/61 and Makassar 757 (the propagating strain for El Tor phages) were found to be stably non-haemolytic. Since the formulation of the Greig haemolytic test (1914), various modifications have been suggested for obtaining uniform and consistent results. A detailed study of the experimental variables has been made by Feeley and Pittman (1963). This method gave reproducible results and was sensitive enough to demonstrate also traces of haemolysin. However, even with this test, not all strains of El Tor vibrios were haemolytic. Barua and Gomez (1967) in a systematic study of 2000 strains of El Tor vibrios examined isolates from the Philippines between 1961 and 1965. They applied various tests including Feeley and Pittman's (1963) as well as Barua and Mukherjee (1964) haemolysis tests and found that all strains isolated in 1961 were haemolytic to sheep erythrocytes when tested soon after isolation. Some non-haemolytic strains were isolated in 1963, and the number increased in 1963, although they showed their haemolytic character by the Barua and Mukherjee's modification of the test. De Moor (1963) reported El Tor vibrio strains isolated in West New Guinea cholera epidemics of 1962–63 to be stably non-haemolytic. In fact, in contrast with the findings of earlier authors, most of the El Tor type of *V. cholerae* strains recently isolated in India are either feebly haemolytic or apparently or truly non-haemolytic, rather than frankly haemolytic. This deviation in the haemolytic character of El Tor strains has often caused considerable confusion about their true identity. A number of tests have been developed for the purpose of differentiation, the more useful of which are the chicken erythrocytes agglutination test (Finkelstein and Mukerjee, 1963) and the Polymyxin B sensitivity test (Gan and Tjia, 1963). The use of differences in phage susceptibility patterns for differentiating the two O group I vibrios was reported by Mukerjee (1960, 1963a, b), Mukerjee and Guha Roy (1961a, b), and Newman and Eisenstark (1964). The latter authors observed that distinct differentiation could be obtained on the basis of their sensitivity patterns for a number of bacteriophages. They, therefore, concluded that the test was promising but it has not been applied to a large series of strains.

The test developed in our laboratory was based on the observation that a group IV cholera typing phage at its routine test dilution is constantly lytic for all classical *V. cholerae* strains but for none of the El Tor strains. The results obtained after testing several thousands of strains with the comparatively dependable tests available are, for comparison, given in Table II.

TABLE II

| Test | Biotype | Total No. of tests | No. showing variation in the expected results | Percentage of variation |
|------|---------|--------------------|-----------------------------------------------|-------------------------|
| 1. Haemolysis of sheep erythrocytes | classical | 331 | 0 | 0·0 |
|  | *eltor* | 1,085 | 443 | 40·8 |
| 2. Agglutination of chicken erythrocytes (GCA) | classical | 331 | 1 | 0·3 |
|  | *eltor* | 879 | 8 | 0·9 |
| 3. Polymyxin B | classical | 587 | 0 | 0·0 |
|  | *eltor* | 585 | 19 | 3·2 |
| 4. Phage susceptibility | classical | 5,693 | 0 | 0·0 |
|  | *eltor* | 1,090 | 0 | 0·0 |

In an earlier study Mukerjee (1963b) reported similar results showing superiority of the phage susceptibility test over the soda-serum agglutination and soda-sublimate precipitation tests of Tanamal and Gipsen and Mayer's tests of chloroform and heat inactivation of agglutinability of *V. cholerae*.

It would therefore appear that of all the tests developed for differentiation of the two biotypes of O group I vibrios, phage susceptibility gives the most consistent results while the results of Polymyxin B sensitivity and agglutination of chicken erythrocytes (GCA) run closely parallel. The Polymyxin B test has been modified by Roy *et al.* (1965) to a simpler and more dependable technique than the original test. The chicken erythrocyte agglutination of Finkelstein and Mukerjee has also been found quite reliable with freshly isolated strains.

In recent years, many strains have been isolated which were non-haemolytic by Feeley and Pittman's modified test, but more like *eltor* biotype *V. cholerae* in other aspects (de Moor, 1963; Roy and Mukerjee, 1966; Barua and Gomez, 1967). Many of these strains were, however, observed by Barua and Gomez (*loc. cit.*) to be haemolytic by the method of Barua and Mukherjee (1964) where 1% glycerol is added as stabiliser (Liu, 1959) in the heart infusion broth for culture of the vibrio strain.

Barua and Gomez (1967) reporting their observations on tests commonly employed for the characterisation of El Tor vibrios found that cholera group IV phage sensitivity gave uniformly good results. Variations in the results of haemolysis tests performed by different methods were observed by various investigators (Tanamal, 1959; Wahba and Takla, 1962; Roy and Mukerjee, 1966; Feeley and Pittman, *loc. cit.;* de Moor, *loc. cit.;* Felsenfeld, 1964; Barua and Mukherjee, *loc. cit.;* Barua and Gomez, *loc. cit.*). One strain of *eltor* biotype of *V. cholerae* which may appear non-haemolytic by the conventional haemolysis test, may be haemolytic by a modification. Our observations also revealed wide variations in results by different methods of haemolysis test. Addition of glycerol to heart infusion broth for culture of the vibrios proved more strains to be haemolytic. A fair number of *V. cholerae* strains isolated during the past few years were originally identified as the *eltor* biotype by tests including phage IV insensitivity, but proved non-haemolytic. On retesting them with glycerol added to the culture media, all these strains proved haemolytic.

Very few reports of irregularities in the expected results of this test (phage IV sensitivity) have so far been published. Monsur *et al.* (1965) reported that group IV phage in sufficient titre on lawns of El Tor vibrios will produce clearings resembling phage lysis. In fact, for satisfactory results the phage has been recommended to be used at RTD as determined on the standard *V. cholerae* strain No. 154. Phage IV resistant classical *V. cholerae* strains have been found in laboratory stocks which retained all other characteristics of the parent strain (Nobechi, 1964; Rizvi and Benenson, 1966). But Barua, (1974), while reporting these observations, has submitted that – "These rare observations, however, have not materially affected the usefulness of the technique for differentiating the strains of classical *V. cholerae* from El Tor vibrios".

Barua and Gomez (1967) recorded their observations on the tests commonly employed for the characterisation of El Tor vibrios using 2000 strains isolated in the Philippines between 1961–1963. The comparative results obtained with different tests are summarised here.

## A. Haemolysis test

Using Feeley and Pittman's method, some strains appeared non-haemo-lytic. But by the technique of Barua and Mukherjee (1964), all were haemolytic. The authors concluded that "these observations indicate that the vibrio that invaded the Philippines in 1961 had typical El Tor properties including that of being haemolytic, and became non-heamolytic or weakly haemolytic in the course of years . . .". This weakly haemolytic property of strains of the Philippines could be demonstrated by growing the organisms in heart infusion broth containing 1% glycerol as stabiliser (Liu, 1959). It may be noted that even this medium was not suitable for detection of all strains isolated in 1964 and 1965.

## B. Agglutination of chicken erythrocytes (GCA)

No irregularity occurred in this test on freshly isolated strains, except 4 "atypical strains", which were non-haemagglutinating and sensitive to Polymyxin B, but resistant to group IV phage.

## C. Polymyxin B sensitivity

Irregular unexpected reactions were observed in the test following the method of Gan and Tjia (1963) which depended on the age of the broth culture used for the lawn and also on the pH of the agar media in the plates.

## D. Cholera group IV phage sensitivity

This test gave uniformly good results except with the 4 atypical strains.

It would thus appear that all the four so called "infraspecific" characters for differentiating the two biotypes of *V. cholerae* run in parallel and that the phage IV sensitivity test is the most dependable of the series.

## V. SEROLOGY OF *V. CHOLERAE*

The specificity of agglutination of bacteria with antisera was first demonstrated by Gruber and Durham in 1896 using the cholera vibrio and typhoid bacillus. Since then, the simple test of slide agglutination has proved sufficient to identify colonies suspected of being cholera vibrio. The *V. cholerae* agglutination scheme is much simpler than that for *Salmonella*.

The presently recognised serotyping scheme of the species (Fig. 1) was laid down in essence by Gardner and Venkatraman in their classic work more than 40 years ago (Gardner and Venkatraman, 1935). After a great deal of controversy (reviewed by Pollitzer, 1959), and more recently after the standardisation of bacteriological criteria, the serotyping scheme has been confirmed and enlarged.

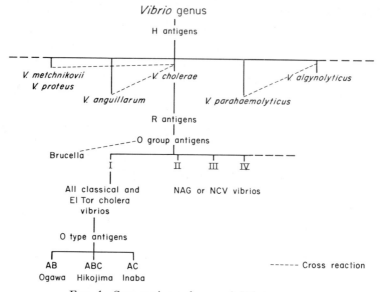

FIG. 1. Sero-typing scheme of *Vibrio* genus.

This scheme is based upon the agglutination reaction for detection of O and H antigens. A "mucoid" M antigen has been described (Sakazaki *et al.*, 1970), but it is rare and a poor antigen. Accordingly, it is of practical importance only because it inhibits agglutination. Although *V. cholerae* is sparsely fimbriated (Barua and Chatterjee, 1964), nothing is known of the fimbrial serology.

The heat-sensitive flagellar antigens of *V. cholerae* and possibly of all vibrios differ from those of other genera in that they are now known to be serologically uniform and definitive of the species (Bhattacharyya *et al.*, 1970; Bhattacharyya and Mukerjee, 1974; Bhattacharyya, 1975; Sakazaki, 1968, 1970; Hugh and Feeley, 1972). They may usefully be applied to identification at the species level. Contrary to common opinion, the flagella are good antigens and H agglutination under suitable conditions is sensitive, reactions occurring at high serum dilutions (1/40,000 and above). Interestingly, the H antigens are not involved in the immune immobilisation reaction. This reaction involves the type-specific O determinants which may also be situated on the complex flagellar structure of vibrios. Relationships of possible phylogenetic importance between minor H determinants of *V. cholerae* and *V. anguillarum*, *V. metchnikovii* and *V. proteus* have been demonstrated (Bhattacharyya, 1975).

Sil *et al.* (1972) in a comparative study of the antigenic composition of agglutinable and non-agglutinable vibrios by gel diffusion and intra-gel absorption tests reported that one O antigen is common to all non-agglu-tinable and O group I *V. cholerae* strains tested. More than one antigen fraction was common to all or some of the agglutinable and some of the NAG vibrios. Each NAG strain tested had its own specific antigens.

*V. cholerae* is now serologically typable into over 50 groups on the basis of the smooth O group agglutinins (Sakazaki *et al.*, 1970; Sakazaki and Shimada, 1975). All major epidemics and pandemics of cholera so far have been caused by strains of the same O group, designated O group I, although other O groups (NCV and NAG vibrios) are being encountered increasingly in outbreaks of cholera-like diarrhoeae. They are always found in the intestinal flora, natural waters, or the stools of patients with O group I cholera and may be genetically inter-convertible with the O group I serotype under experimental conditions (Bhaskaran and Sinha, 1971). A positive reaction of suspect colonies with O group I antisera is considered to be diagnostic of cholera.

Strains in the rough form are frequently encountered at the tail end of epidemics and in chronic carriers. Rough O determinants are anti-genically poor and are not group specific. Considerable inter-group antigenic similarity exists (White, 1935; Shimada and Sakazaki, 1973) and poses a difficult serodiagnostic problem. The rough determinants may be related to the species-wide O agglutinins which appear after heat-de-naturation (Gardner and Venkatraman, 1935; Lyles and Gardner, 1958). Sanyal *et al.* (1972) reported the possibility of R (Rough) strains being identified as derived from biotypes of O group I *V. cholerae* by phage-typing.

Interactions between the O antigens of cholera vibrios and many species including most intestinal bacteria have been reported (Pollitzer, 1959; Feeley, 1969; Gangarosa *et al.*, 1970; Barua and Watanabe, 1972; Winkle *et al.*, 1972). The only firmly established cross-reaction is with *Brucella* (Feeley, 1969). Whilst this may cause confusion in interpreting the sig-nificance of antibodies in a given patient, it does not interfere with the serotyping of cultures.

The serological infra-structure of the O group I vibrios still remains a subject of controversy. Smooth strains isolated in the field are routinely classified into three O types, Ogawa, Inaba and the intermediate Hikojima, the names denoting their historical origin. Type specificity is culturally quite stable. Polyvalent antisera raised against any one of these types will react with all three because of a common group factor designated A. Mono-specific Inaba and Ogawa O sera are prepared by cross-absorption of polyvalent sera. These reactions imply a minimum of three O determinants, the group factor A, and the type-specific factors B (Ogawa) and C (Inaba). The existence of the Inaba C factor has been questioned on serological

(Kauffmann, 1950; Sakazaki and Tamura, 1971) and immunochemical (Redmond *et al.*, 1973) grounds. On this basis, two antigenic factors, A and B, are of major significance, other antigenic differences being quantitative rather than qualitative. The two-determinant structure implies that the Hikojima and the Inaba types arise from Ogawa by loss of the B factor, they may therefore more appropriately be considered as variants than types. This unidirectional mutation Ogawa-Inaba-Rough can readily be demonstrated to occur *in vitro* at a frequency of $1 \times 10^{-7}$ by exposure of *V. cholerae* to type-specific O antisera (Bhaskaran and Gorill, 1957; Sakazaki and Tamura, 1971a, b). *In vivo*, however, multidirectional serotype changes associated with the development of immunity have been demonstrated in gnotobiotic animals (Sack and Miller, 1969). The three-determinant schemes would therefore appear to be epidemiologically valid and of diagnostic utility, although no Inaba to Ogawa variation in epidemics is on record.

In view of the uncertainty of the analytical serology of *V. cholerae* it is not surprising that the immunochemical basis of O specificity is not yet understood. It seems certain that as in the *Enterobacteriaceae*, the O agglutinogens are located in the cell wall protein lipopolysaccharide complex (Misra *et al.*, 1961; Gallut, 1962; Pant, 1968). These antigens are also the target of vibriocidal antibody activity (Neoh and Rowley, 1970). The presence of this complex in culture supernatants or other cell fractions probably reflects an autolytic or secretory process. Factors common to the species as well as a part of the O group I A factor are probably carried by the protein moiety (Gallut, 1965; Redmond, 1975). The specificity of type factors B and C and also the remainder of the group A factor is thought to be determined by the cell wall lipopolysaccharide. The lipopolysaccharide of *V. cholerae* differs from that of *Enterobacteriaceae* in that it lacks the core component 2-keto-3-deoxy-octonate (Jackson and Redmond, 1971; Jann *et al.*, 1973). Apart from only quantitative differences of the principal sugars, glucose, glucosamine and heptose in the different O groups (Shrivastava, 1965), substantial quantities of unusual amino-sugars have been found, the instability of which has prevented their detection until recently (Jackson and Redmond, 1971; Redmond *et al.*, 1973; Jann *et al.*, 1973; Redmond, 1975). The importance of these amino-sugars and also O-acetyl groups (Guhathakurta and Dutta, 1974) in determining O-specificity remains to be established. Type-specific Inaba determinants have been demonstrated in the cell wall (Ghosh and Mukerjee, 1961; Mukerjee, 1963a, b) and lipopolysaccharide (Misra *et al.*, 1961; Pant, 1968; Finkelstein, 1962) of O group I vibrios.

Serotyping of *V. cholerae* is normally performed by agglutination of cell suspensions on a glass slide or in tubes. However, rapid methods of identification and typing of vibrios directly in stool samples by fluorescent antibody labelling (Finkelstein and La Brec, 1959) or immobilisation (Benenson *et al.*, 1964) with group and type specific O antisera have been successfully applied in the field on a limited scale. Retrospective analysis of serum samples from bacteriologically negative cases or contacts would have been epidemiologically useful, but the overlapping titres of antibody in patients and normal persons in endemic areas precludes this.

Ghosh and Mukerjee (1961) using gel diffusion and intragel absorption of lysates of agglutinable and 26 NAG vibrios found that antigenic relationships exist between them. NAG vibrios possess 2 to 4 heat-labile antigens common with those of *V. cholerae* (classical). The NAG vibrios possess one non-specific heat-stable somatic antigen identical with that of *V. cholerae* (classical). Like the Ogawa and Inaba serotypes, the NAG strains exhibit their own specific heat-stable precipitation band, which differs from strain to strain.

Attempts to subtype the cholera vibrios by analysis of intracellular (Misra and Shrivastava, 1961) and extracellular antigens have been unsuccessful. However, Vasil *et al.* (1974) have found that the *eltor* biotype did not react with antitoxic sera in Elek plates, whereas the classical biotype did. The significance of this finding is yet to be assessed.

## VI. SEROTYPING METHODS

A survey of the literature indicates that the techniques used are diverse. A standard method has yet to be adopted. Rather than attempt to be comprehensive, I have based this account on common practice adding the experience gained in our own laboratory. Most techniques have been developed with the O group I cholera vibrio in view, but there is every indication that they are equally applicable to both the NAG vibrios, and other vibrio species (Sakazaki *et al.*, 1968; Bhattacharyya, 1975).

Most of the needs of the clinical cholera worker can be satisfied by reference strains of Inaba and Ogawa O group I *V. cholerae* and their O antisera. Polyvalent O antiserum is available commercially and is also easy to prepare. The type-specific O antisera prepared by absorption, are sufficient to serotype most clinically encountered strains; the more difficult isolates may be referred to one of the Vibrio Reference Centres.

However, to be able to identify and serotype vibrios of the other O groups (the NAG or NCV vibrios) on the basis of the antigenic structure as presently understood (Fig. 1), four kinds of diagnostic antisera and reference strains are required.

### 1. *Species identification with H antisera*

The results of Bhattacharyya (1975) and Sakazaki and Shimada (1975) indicate that H sera can be used to identify all O groups of *V. cholerae* more rapidly and conveniently than by bacteriological tests.

### 2. *Sero-grouping with O antisera*

Cultures reacting with H antisera can be grouped by O antisera prepared against the 53 reference strains described by Sakazaki and Shimada (1975).

### 3. *Typing sera of O group I* V. cholerae

The two typing sera, Inaba and Ogawa specific O antisera, are used to type O group I *V. cholerae;* the intermediate Hikojima reacts with both. The type characteristics of other O groups are not yet known.

### 4. *Rough O antiserum*

Strains which react with H sera, but not with O grouping antisera, may be in the rough form and can be identified with R serum.

Apart from the O grouping antisera, all these sera can be prepared from reference strains of O group I *V. cholerae.*

## A. Preparation of cultures

### 1. *Maintenance of cultures*

Because of high mutation rate (Bhaskaran and Gorill, 1957) *V. cholerae* in culture tends to dissociate to immobile, rough, rugose, or mucoid forms (Pollitzer, 1959). Lyophilisation is the only reliable means of preserving reference cultures. However, this is not always feasible and we have maintained a large collection of very old strains at ambient temperature by monthly subculture on nutrient agar. *V. cholerae* also can be maintained in stab culture, but such subcultures do not survive well in the refrigerator.

### 2. *Isolation of smooth form strains from apparently rough cultures*

It is important to prepare O group and type antisera from smooth cultures to avoid high R titres which are not group specific. Dissociation to roughness is not always obvious from culture characteristics and can be detected by agglutination with R antiserum or with 0·2% trypaflavin. R dissociants, which are culturally stable, are occasionally encountered in chronic carriers and convalescent excretors. The rough or partially rough cholera vibrio is sensitive to complement and can be converted to the smooth form by one of the following procedures:

(a) A loopful of a 3-hours broth culture is inoculated into nutrient broth containing 5% complement and plated after growth at 37°C for 18 hours. Smooth colonies are selected by agglutination with O group antisera and should be checked with R serum or trypaflavin.

(b) A guinea-pig is injected intraperitoneally with 1 ml of a 3-hours broth culture diluted 1 in $10^3$. After 18 hours, the peritoneal exudate is plated and smooth colonies selected. It may be necessary to repeat these procedures for very rough strains.

### 3. *Isolation of flagellated forms*

Although freshly isolated vibrios are actively motile, motility is rapidly lost in culture. For the production of H antiserum and use as antigen in H agglutination it is important to select motile cell lines from stock or reference cultures. This is most simply done by serial passage in nutrient Craigie tubes containing 0·4% agar and a little surface broth. Motility is stable in lyophilised cultures.

### 4. *Isolation of rough variants*

By incorporating 0·3% of complement inactivated O group antisera in Craigie tubes, Shimada and Sakazaki (1973) could isolate rough variants from most O groups of *V. cholerae*. The results of White (1935) indicate that the use of a minimum effective serum concentration is important to avoid the induction of involution forms by a low level R antibody present in O group antisera.

### 5. *Other variants of* V. cholerae

Although strains freshly isolated from patients or the environment are usually motile and smooth, and present no problems, rugose variants have been encountered in chronic carriers. The rugose variant is typically adherent to agar, but carries specific O antigens which may be determined if it can be brought into suspension. It also dissociates to form smooth colonies on plating which can be recognised morphologically (Azurin *et al.*, 1967). Sakazaki *et al.* (1970) have also described a mucoid or M form which does not agglutinate with O antisera, although it is immunogenic. Mucoid cultures become O agglutinable after boiling for 2 hours and should be grouped and typed by tube agglutination.

## B. Production of antisera

Young adult rabbits with pre-immunisation titres to whole-cell (OM) antigens of less than 1/50 are used for production of antisera. *V. cholerae* antigens are toxic to rabbits, particularly between the 20th and 30th days of the immunisation schedule (Bhattacharyya and Mukerjee, 1974). Fatalities can be avoided by increasing the intervals between injections.

## 1. *Polyvalent O group antisera*

To obtain group specific O antisera free of H antibody, boiled cells are used as antigens. It is recommended to boil saline suspensions of cells grown at 37°C overnight in a water bath for 3 hours or under a reflux condenser for 2 hours. Alternatively, the cell suspension may be auto-claved at 15 lb for 20 min. Thereafter the cells are washed thrice in saline, and the suspension standardised to an optical density of 0·8 at 540 nm (equivalent to a Klett-reading of 200 or a viable count of $3 \times 10^2$ cells/l). The suspension is injected intravenously in graded doses from 0·5 to 2·0 ml twice weekly. Satisfactory O titres (1/2000 to 1/5000) are usually obtained 7–8 days after the 6th injection. The animals may be exsanguinated by cardiac puncture, or bled periodically from the ear vein.

The antiserum (whether prepared by the worker himself or obtained from commercial or other sources) should be checked for specificity. Tested with a sensitive OH preparation of another O group, its H titre should be below 1/100. From the data of Shimada and Sakazaki (1973), it would seem that the R activity should not be greater than 1/20. If neces-sary, the antiserum should be absorbed with appropriate cell suspensions. The antiserum should also be free from reactions with common intestinal pathogens. The faulty diagnosis of a ship outbreak of salmonellosis as cholera (Gielen *et al.*, 1972) is a case in point.

## 2. *Monospecific O typing sera for O group I vibrios*

Polyvalent O group I antisera raised against reference strains of the Inaba and Ogawa types are made monospecific by absorbing out the common A determinants. This is usually done by reacting diluted (1/10) O serum with approximately one-tenth its volume of packed cells of the heterologous type at 37°C for 4 hours and then at 4°C overnight. The O titre to the absorbing antigen usually drops below 1/20 after performing the absorption thrice. Sakazaki and Tamura (1971) demonstrated that Inaba activity could be completely removed by repeated absorption. Compared with Ogawa activity, Inaba activity is weak and it seems probable that this loss of activity is due to non-specific over-absorption and it should be avoided. The column method of absorption described in section B 4 (*vide infra*) for H sera may also be used to prepare high titre O sera.

The usual practice of using live vibrios to absorb antisera is not advisable, because the bacteria are proteolytic. Heat-inactivated (56°C for 1/2 hours) or formolised (0·3%) cells are suitable.

### 3. *Rough O antisera*

Shimada and Sakazaki (1973) obtained R antisera from most O groups of *V. cholerae* by injecting 8 or 9 doses of heat-denatured rough cells. R antisera should not have O titres greater than 1/20.

### 4. *H antisera*

Bhattacharyya and Mukerjee (1974) found that H titres of 1/40,000 and above were obtained after 8 or 9 injections of formalin fixed (0·3%) suspensions of motile vibrio isolates. O activity was removed by absorption with autoclaved O antigen dispersed in an inert matrix of cellulose in a chromatography column. The OH serum was recycled through the column over a period of 4–5 days in the cold. O antigen from two nutrient agar plates in 16 g (wet weight) of washed cellulose powder was adequate for the absorption of 2 ml of serum in a volume of 20 ml. This method was chosen to avoid the 5- to 10-fold non-specific loss of H activity caused by the conventional procedure described above. The O titre should fall below 1/20.

### 5. *Preservation of antisera*

Although diagnostic typing antisera are best maintained with added preservatives (merthiolate 0·01%, sodium azide 0·1%, or phenol 0·25%), it is advisable to maintain sterile aliquots without antibacterials for possible future use with living systems, e.g. for the isolation of rough variants. Antisera should be divided into small aliquots and stored in ampoules or screw-capped tubes at 4°C or frozen at −20°C or in the lyophilised state.

## C. Standardisation of antisera by tube agglutination methods

*V. cholerae* antisera are usually titrated by mixing equal volumes (0·2 to 0·5 ml) antigen suspension of standard density and serum dilutions in agglutination tubes. Strains vary in their intrinsic sensitivity to agglutination; four-fold titre differences may be accepted as within normal limits. It is important to control assays with preparations which are known to react well. Opposing views on most aspects of *V. cholerae* serology have been reported because optimal testing conditions have not been used.

### 1. *O agglutination*

The technique of O agglutination includes using (1) live suspensions of fresh cultures and (2) an incubation temperature of 37°C. The method of De Lahiri *et al.* (1958) is most widely used. Overnight nutrient agar culture (37°C) is suspended in normal saline to an opacity equivalent to a broth culture ($10^8$–$10^9$ cells/ml) and reacted with serum dilutions at 37°C for 4 hours. Final readings are taken after holding the tubes in the refrigerator overnight. Barua and Sack (1964) demonstrated satisfactory O agglutination in very dilute 3-hours peptone water cultures by incubation for 1 hour before refrigeration. Clearing of the tubes to titre was evident at the end of incubation.

These tests are somewhat inconvenient because they require freshly prepared live suspensions where the cells are variable and tend to autolyse. Fixing the suspension slows down, but does not inhibit the O reaction. Maximal O titres were obtained with formolised suspensions by prolonging the incubation to 18 hours (Guha Roy and Mukerjee, 1959). Heat treated suspensions are unsatisfactory as agglutinogens. Four-fold reductions in agglutinability are seen after even brief heat treatment (Guha Roy and Mukerjee, 1959). However, it is necessary to boil mucoid variants which do not otherwise react with O antisera (Sakazaki et al., 1970). In view of the finding that R determinants become reactive after prolonged heating (Lyles and Gardner, 1958), it is important in these tests that the antisera should be free of R activity.

In view of possible interference by the H components at dilutions near the H equivalence point of 1/300 (Bhattacharyya, 1970) estimates of O activity in OH antisera by these methods are ambiguous. It is possible to estimate the O titre of OH sera by using suspensions of vibrio cells which have been deflagellated by homogenisation in a tissue grinder at 16,000 r.p.m. for 2 minutes followed by centrifuging and washing (Bhattacharyya and Mukerjee, 1974).

## 2. H agglutination

The H reaction is slow. Optimal conditions are the use of (1) dense formolised suspensions (Klett 400 at 540 nm) of motile isolates and (2) a high reaction temperature (52°C for 18 hours).

It is difficult to demonstrate H activity with dilute suspensions or with living or merthiolated suspensions at low incubation temperatures. This may be due to the limited number of antigenic sites available in monoflagellar cells and degradation of the flagella occurring early after autolysis. These reasons may explain why Smith (1974) was unable to demonstrate an H relationship between the different O groups of V. cholerae.

The H activity of OH antisera although normally well beyond the range of O activity, can be estimated by using suspensions of other O groups.

A 2-hours diagnostic test for NAG and cholera vibrios using H serum dilutions at the H equivalence point has been described (Bhattacharyya, 1975).

### 3. *Rough O agglutination*

Rough cultures are suspended in 0·4% saline to avoid auto-agglutination. Shimada and Sakazaki (1973) used suspensions which had been boiled for 2 hours and incubated at 50°C for 18 hours. In view of the low titres (1/160) of these authors compared with the high titres of White (1935) who employed live suspensions, it would seem that the optimal conditions for the R reaction are not yet understood fully.

### D. Typing by slide agglutination methods

For most typing purposes, the slide agglutination test is adequate. The antisera must be standardised and checked for specificity by tube agglutination methods.

Overnight nutrient agar cultures (37°C) are suspended evenly in normal saline and mixed with serum dilutions by tilting the slide. It is not advisable to rely on positive reactions from colonies on selective media. TCBS agar for example may give rise to false positive tests. Auto-agglutination of rough strains may take place only slowly, if at all. Such strains should be suspended in distilled water. The optimal working dilution of antiserum should be determined with known homologous strains. High serum concentrations (dilutions of 1/5 or less) are liable to show non-specific or prozone effects. With good antiserum preparations the actual working dilutions are about 1/25 for O grouping or H antisera and 1/15 for O type sera. Group O and Ogawa type O reactions are strong and immediate. Inaba type O, H and R reactions are slower and weak. H reactions should be read after 30 seconds.

### E. Other typing methods

Several serological techniques have been developed which may be of value in particular situations. With retrospective diagnosis in mind, the O agglutination reaction has been modified for the handling of small blood samples on a large scale. Benenson *et al.* (1968) miniaturised the test for use with disposable kits. Barua (1963) described a capillary tube and microscope agglutination test. Microscope methods have also been applied to NAG vibrios (Givental *et al.*, 1974). These techniques are as sensitive as macro-agglutination tests. Vibriocidal antibody assay methods are a hundred-fold more sensitive. Applied to the assay of serum antibody levels, these methods have group, but not type, specificity, although it has been possible to identify type specific antibodies in the vibriocidal antibody inhibition test by using purified somatic antigens (Finkelstein, 1962).

Microscopic methods which have been used successfully in the field for rapid identification and typing of cholera vibrios by direct application of

O antisera to stool samples, or their 2-hours enrichment cultures, are dark field immobilisation (Benenson *et al.*, 1964) and fluorescent antibody labelling (Finkelstein and La Brec, 1959). These techniques require a vibrio count of approximately $10^6$ per ml.

## VII. PHAGE TYPING OF *V. CHOLERAE*

### A. Introduction

Although Asheshov *et al.* (1933) recorded a difference in the sensitivity of *V. cholerae* strains to bacteriophages, systematic studies were not made until 1955 when in accordance with a recommendation of the WHO Expert Committee (Report 1952) on Cholera, a bacteriophage typing scheme for *V. cholerae* (classical biotype) was developed in the Indian Institute of Biochemistry and Experimental Medicine, Calcutta. This was widely applied to epidemiological investigations during 1955–1965 and yielded much valuable information (Mukerjee, 1967; 1973; 1974).

Attempts to develop potentially important phage typing schemes for *V. cholerae* were those of Nicolle and colleagues (Pasteur Institute, Paris, France), Mukerjee and his associates (Indian Institute of Biochemistry and Experimental Medicine, Calcutta, India), Newman and Eisenstark (Kansas University, Manhattan, U.S.A.), and Takeya and his co-workers (Kyushu University, Fukuoka, Japan). In spite of valuable contributions by Nicolle *et al.* (1960; 1962) and Newman and Eisenstark (1964), their work probably suffered from the limitations that only a small number of phage and vibrio isolates (old) were available to them.

### B. Phage typing for strain identification of *V. cholerae*

Working in an endemic area (Calcutta), Mukerjee and co-workers could test over 6000 fresh faecal isolates of *V. cholerae* and many phages in the light of their epidemiological background and produce a fully-fledged typing scheme. The functioning of an International Reference Centre for Vibrio phage typing (WHO) in Calcutta since 1963 under the direction of the author facilitated further development of phage typing schemes due to international co-operation and a plentiful inflow of current vibrio isolates for reference purposes.

Studies conducted by the author and his co-workers since 1955 resulted in the development of two separate phage typing schemes for the classical and *eltor* biotypes of *V. cholerae;* these enabled identification of strains with a high degree of precision. Subsequently Sil *et al.* (1974) developed a phage typing scheme for the O : I negative vibrios (NAG vibrios). These different phage typing schemes are described below.

## C. Phage typing of the classical biotype of *V. cholerae*

Nicolle and his collaborators (1960; 1962; Gallut and Nicolle, 1963) grouped *V. cholerae* strains on the basis of their sensitivity patterns to eight lytic phages from various sources. Newman and Eisenstark (1964) studied a comparatively larger series of strains against different groups of phages. According to Nicolle *et al.* (1962), there were four phage types among the 1959–60 Bangkok epidemic strains of "classical" *V. cholerae* and two types among three biotype *eltor* strains and 34 NAG vibrios, that were common to the classical type. Working with these phages, we found that strains originating from a single source of infection belonged to more than one type, while *V. cholerae* (classical) isolated from the Bangkok epidemic belonged to a single phage type (Type 1); the strains of *V. cholerae* biotype *eltor* differed in phage types from *V. cholerae* (classical) and the NAG vibrios showed susceptibility patterns different from those of the two O group I vibrios (Mukerjee and Guha Roy, 1961a, b).

In our studies, five principal phage types of *V. cholerae* (classical biotype) have been identified (Mukerjee *et al.*, 1957, 1959, 1960; Mukerjee, 1963a, b) on the basis of their sensitivity patterns to the four groups of cholera bacteriophages isolated from stools of cholera patients (Fig. 2). One of these types (type 2) could be further subdivided into three subtypes by adapted phages (Mukerjee *et al.*, 1957). Another type (type 1) was subsequently divided into three subtypes by their susceptibility to new bacteriophages isolated from a lysogenic strain of *V. cholerae* originating in Bangalore (Mukerjee, 1965).

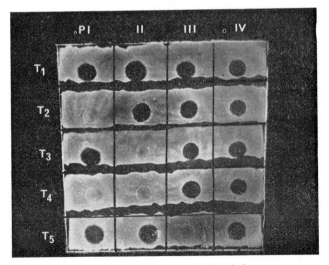

Fig. 2. Five phage types of *V. cholerae*.

## D. Phage typing of *eltor* biotype strains of *V. cholerae*

Phage typing of *V. cholerae* biotype *eltor* has been described by Nicolle *et al.* (1960), Gallut and Nicolle (1963), Newman and Eisenstark (1964), and Mukerjee (1965). As Gallut and Nicolle examined a total of only 14 strains of biotype *eltor* and their scheme provided only a broad classification into 3 groups, its practical value is limited (Mukerjee, 1965). The value of the scheme of Newman and Eisenstark is yet to be assessed. The scheme initially proposed by Mukerjee (1965) had to be discarded soon afterwards due to anomalous results.

A phage typing scheme using five groups of typing phages was adopted by which six phage types could be distinguished among *V. cholerae* biotype *eltor* strains (Basu and Mukerjee, 1968) (Table III).

TABLE III

**Phage typing scheme for *V. cholerae* biotype *eltor*—1968**

| Phage type of *V. cholerae* biotype *eltor* | Lysis by phage group | | | | |
|:---:|:---:|:---:|:---:|:---:|:---:|
| | I | II | III | IV | V |
| 1 | + | + | + | + | + |
| 2 | + | + | + | − | + |
| 3 | + | + | − | + | + |
| 4 | + | + | − | − | + |
| 5 | + | − | − | − | + |
| 6 | − | + | − | − | + |

The above scheme has been in general use and proved to be sound. The chief utility of phage typing for both the biotypes of *V. cholerae* is the tracking of epidemics from one place to another, rather than tracing chronic carriers or spread of infection through individuals.

## E. Phage typing of *V. cholerae* other than group O sub-group I *V. cholerae* (non-agglutinable or NAG vibrios)

In view of the recent reports that NAG vibrios are choleragenic it was of interest to develop an epidemiologically useful phage typing scheme for the NAG vibrios and attempts have been made in our laboratory to do this. A tentative scheme was proposed (Table IV) (Sil *et al.*, 1974), but further work is necessary to correlate the phage types with serotypes and to establish epidemiological utility (Sil *et al.*, 1972).

## F. Evaluation of phage typing for epidemiological studies and documentation of the results

Studies on phage typing of *V. cholerae* for epidemiological purposes

## TABLE IV

**Phage sensitivity patterns of NAG vibrio strains. NAG phage typing pattern**

| Bacteriophages Lab. No. (NAG typing phage no.) | Bacteriophage sensitivity patterns | | | | | | | | | | | | | |
|---|---|---|---|---|---|---|---|---|---|---|---|---|---|---|
| | 1 | 2 | 3 | 4 | 5 | 6 | 7 | 8 | 9 | 10 | 11 | 12 | | * |
| NAG phage no. 09 (I) | + | − | + | − | − | + | + | − | − | − | − | + | − | + |
| NAG phage no. 29 (II) | + | − | + | − | + | − | + | − | − | + | + | + | + | − |
| NAG phage no. 30 (III) | + | − | − | + | − | − | − | − | + | − | + | − | + | + |
| NAG phage no. 31 (IV) | + | − | − | + | − | − | + | + | − | + | − | − | + | + |
| No. of strains in each phage type | 12 | 62 | 45 | 14 | 19 | 10 | 12 | 7 | 6 | 5 | 2 | 6 | 1 | 1 |
| Percentage of strains in each type | 6·0 | 31·0 | 22·5 | 7·0 | 9·5 | 5·0 | 6·0 | 3·5 | 3·0 | 2·5 | 1·0 | 3·0 | 0·5 | 0·5 |

* = Patterns that have not been numerically designated

carried out in our laboratory during the period 1955–65 showed that some changes have occurred in the characteristics of the bacteria and in the epidemiology of disease. Interested readers may consult Mukerjee (1967; 1973; 1974).

A change of the causative organism of cholera from the classical to the *eltor* biotype has occurred in India (Fig. 3) which formed the original home of cholera. This is of fundamental epidemiological interest. In India we have perhaps been witnessing the last phase of classical cholera,

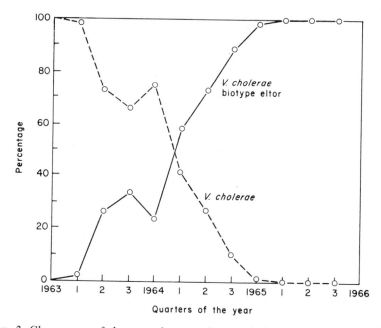

FIG. 3. Changeover of the causative organism of cholera in India (1964–1966).

which has existed in that country since time immemorial. The disappearance of the classical biotype of *V. cholerae* may have been due to a number of factors. These are principally (a) the ability of the *eltor* biotype to eliminate the classical biotype in mixed cultures both *in vivo* and *in vitro* (Basu *et al.*, 1966), (b) the higher carrier rates, (c) the longer duration of the carrier status in El Tor infection and (d) the greater viability of El Tor vibrios in the natural environment. We had foreseen the possibility of change of cholera in India (Mukerjee *et al.*, 1965), but the rapidity with which the change took place was remarkable and hardly has any parallel in the epidemiology of other bacterial diseases.

At least four independent foci harbouring different phage and serotypes of *V. cholerae* were found to be present in the Indo-Pakistan subcontinent (Fig. 4) (Mukerjee, 1965). These foci consisted of (I) East Pakistan and the eastern states of India, viz. West Bengal, Assam, Bihar and Orissa, (II) the South Indian states including Madras, Andhra and Kerala, (III) the state of Maharashtra and (IV) the Gujarat state. In Calcutta, all the five

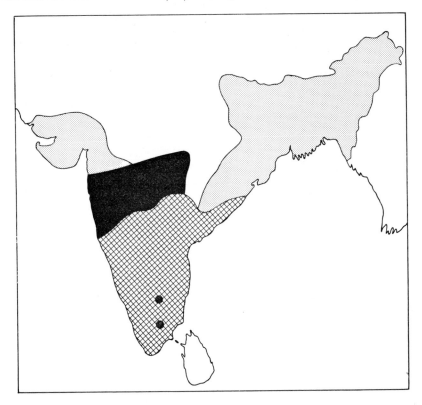

☐ Areas not examined

▨ Areas harbouring type IA vibrios

▨ Areas harbouring cholera endemic foci for type 3 vibrios

▨ Areas harbouring cholera endemic foci for type I and type 3 vibrios

● Areas harbouring type IB and IC vibrios

Cholera phage-typing map of Indo-Pakistan subcontinent

FIG. 4. Cholera phage type map of Indo-Pakistan subcontinent.

phage types of *V. cholerae* (classical) were present periodically between 1955 and 1962. The distribution of the phage types of vibrios in the city did not follow any regular pattern and appeared scattered. Types 1 and 3 were most frequent, while the others were found in small numbers. Until the first quarter of 1960, phage type 1 vibrios formed the overwhelming majority of the strains isolated, while type 3 vibrios were next in order of frequency. A rapid change then took place (Fig. 5) and type 3 Inaba vibrios were soon predominant in Calcutta (Mukerjee, 1963b). Shortly afterwards, other phage types disappeared from the area.

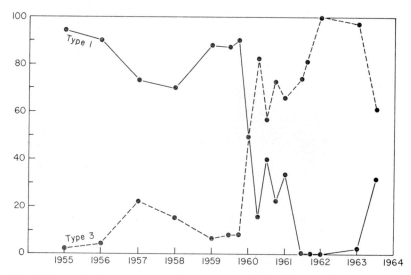

FIG. 5. Variation in proportion of type 1 and 3 *V. cholerae* isolated in Calcutta epidemics.

In 1961–63, El Tor infection came out of its original home in Indonesia and spread rapidly over wide areas comprising the West Pacific Islands and South East Asian countries (Fig. 6). The world incidence of cholera which had been progressively falling during the past few decades, started to show an upward trend. After a fairly extensive investigation on the spot, Felsenfeld (1963) reported that the population shifts in China and troop movements from and to Makassar on the Celebes Island might have been the principal factors contributing to the extensive spread of El Tor in Indonesia. Illegal movement of goods and displaced persons between the countries subsequently involved was thought to be a factor in the widespread dissemination of cholera El Tor outside Indonesia. The spread of infection

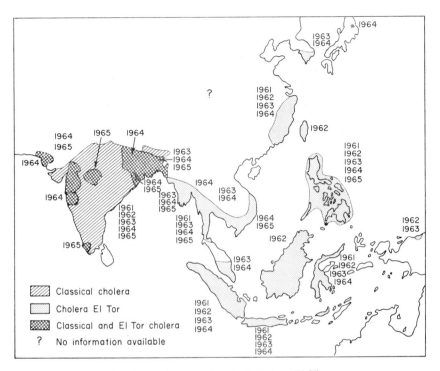

FIG. 6. 1961–1963 spread of cholera El Tor.

took place from Indonesia directly to a number of West Pacific Islands and South East Asian countries. Phage typing indicated that the spread of infection was not always confined to contiguous areas, but sometimes occurred between distant places through commerce, repatriates or immigrants (Mukerjee, 1964a; Mukerjee *et al.*, 1965).

Following the fairly widespread epidemic in Burma in 1963, cholera El Tor appeared in East Pakistan towards the end of that year (Mukerjee, 1964b). In India, El Tor was detected in the Calcutta area in April 1964 (Barua *et al.*, 1964; Mukerjee, 1964a). It was already present in Madras as early as March 1964 (Mukerjee, unpublished data). The disease appears to have entered India at more than one point (Mukerjee *et al.*, *loc. cit.*). The four possible points of entry and the probable lines of spread of cholera El Tor in India have been plotted in the map (Fig. 7) which was drawn on the basis of results of phage typing of El Tor strains isolated in different areas and the chronological sequence of the outbreaks. The entry of cholera El Tor into Calcutta and Madras is likely to have taken place through repatriates from Burma while the infections in Gujarat and the Punjab States may have had their origin in West Pakistan.

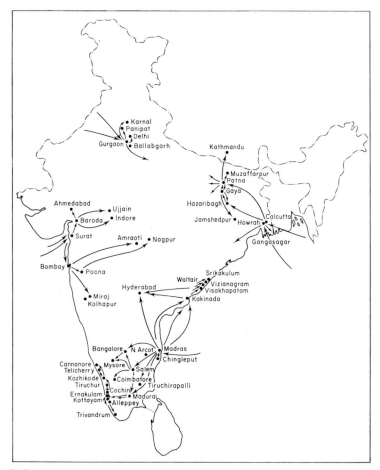

FIG. 7. Entry and spread of cholera El Tor in the Indo-Pakistan subcontinent.

El Tor rapidly spread over India and Pakistan and as anticipated (Mukerjee *et al., loc. cit.*) went beyond the borders of the Indo-Pakistan subcontinent, spreading in 1965 to Nepal, Afghanistan, Iran, Bahrain, and the Southern States of U.S.S.R., and in 1966 to Iraq.

In India, where El Tor infection met pre-existing classical cholera, cholera El Tor rapidly outnumbered and ultimately eliminated cases of the classical type (Fig. 3). Striking evidence for this is that whereas only *V. cholerae* (classical) were isolated in India before March 1964, nearly 89% of about two thousand cholera cultures isolated in 1965 from different parts of India belonged to the El Tor type. In 1966, among the 635 strains from ten Indian states, not a single *V. cholerae* (classical) strain was found. The remaining states were either free from cholera outbreaks, or cholera cultures were not available to us for testing.

The epidemiological validity of the El Tor phage typing scheme has been established by demonstrating the uniformity in phage types of strains in outbreaks originating from single sources of infection. Vibrios isolated from the localised outbreaks in Kowloon in 1964 originated from a single restaurant; similarly outbreaks in Indore, Karnal and Nagpur of India in 1965 and in Korat of Thailand in 1966 originated from a single source of infection. In all these outbreaks, the isolates of *V. cholerae* biotype *eltor* were of identical phage and serotypes. Successive isolates from 14 chronic carriers of *V. cholerae* biotype *eltor* in the Philippines and Thailand, and 37 sets of strains isolated from cholera patients and infections in their known contacts also showed uniformity in phage types and serotypes.

## VIII. BIOLOGICAL PROPERTIES OF *V. CHOLERAE* TYPING BACTERIOPHAGES (PHYSICAL, PHYSICOCHEMICAL AND SPECIAL)

### A. Plaque morphology of cholera typing phages

The plaque morphology of bacteriophages provides a practical means of classification of the phages. The production of different types of plaques by different phages was recognised early by Gratia (1922), Asheshov (1924), Burnet (1933), and Asheshov *et al.* (1933). The last named authors reported detailed observations on the plaque morphology of cholera phages. Observations on the morphology of plaques of different types of *V. cholerae* (classical) typing bacteriophages (Mukerjee, 1961a) are given below. The morphological characteristics of plaques are best seen when phages are plated by the agar layer technique using soft agar (Adams, 1959).

Figures 8a–d show plaques of the 4 groups of typing phages in one set of experiments using a type 1 *V. cholerae* strain, No. 154, as the host bacterium. Of the plaques produced by these four sets of phages, the characters of group II were most distinctive. The group II phages are "smooth specific" lysing only the smooth elements of a culture (Mukerjee, 1959). The rough elements were resistant usually producing secondary growth over the lytic area.

### B. Morphology of cholera typing phages

The preliminary observations of Vieu *et al.* (1965) on the morphology of two of the *V. cholerae* phages of the French workers under the electron microscope after negative staining were reported in 1965. The detailed

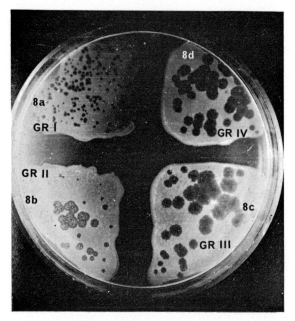

Fɪɢ. 8a, b, c, d. Morphology of plaques of four groups of *V. cholerae* typing phages.

morphological characteristics of the four groups of *V. cholerae* (classical) typing phages have been studied in the Department of Biophysics, School of Tropical Medicines, Calcutta (Chatterjee *et al.*, 1965; Das and Chatterjee, 1966). Negatively stained preparations of the phage particles were made by the method of Brenner and Horne (1959). These investigations confirmed the differences in the four groups of typing phages first detected by their patterns of lysis (Mukerjee *et al.*, 1957) and confirmed later by other characteristics (Mukerjee, 1963a, b). The dimensions of the phages as summarised by Das and Chatterjee (1966) are given in Table V. Typical micrographs are shown in Figs. 9a–d.

Of the five groups of *V. cholerae* biotype *eltor* typing phages, only one phage belonging to group I has been studied, by Maiti and Chatterjee (1969a, b) using the metal shadowing technique. Their investigations revealed that all four groups of *V. cholerae* (classical) typing phages were morphologically distinct (Chatterjee *et al.*, 1965). The dimensions of the *V. cholerae* typing phages and their fine structures as reported by Das and Chatterjee are given in Table V.

Group III phages have a very short tail. Group I phages, on the other hand, apparently have no tail as evidenced by metal shadowing and by

## TABLE V
### Dimensions of cholera phages

| Phage | Head Å | Length Å | Tail Length Å | Knob Å | Prong Å |
|---|---|---|---|---|---|
| Group I | $706 \pm 18 \times 740 \pm 27$ | . . | : : | . . | : : |
| Group II | $612 \pm 31 \times 655 \pm 37$ | $810 \pm 32$ | $166 \pm 20$ | 183 | $156 \times 96$ |
| Group III | $611 \pm 50 \times 644 \pm 53$ | $178 \pm 20$ | $174 \pm 25$ | : : | : : |
| Group IV | $738 \pm 33 \times 336 \pm 40$ | $1528 \pm 82$ | $107 \pm 14$ | 130 | : : |

negative staining. According to Das and Chatterjee, group I phages apparently are adsorbed to host bacteria by the head surface structures. Negative staining in some of the group I phages has revealed the presence of a randomly coiled filament.

## C. Antigenic specificity of *V. cholerae* (classical) typing phages

Bordet and Ciuca (1921) observed that serum of rabbits immunised with

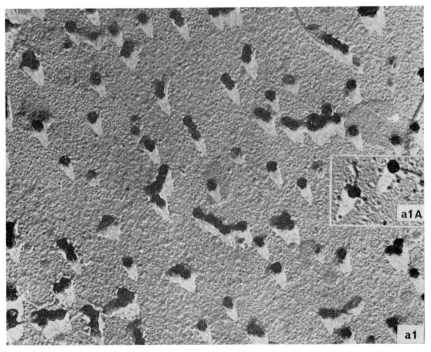

Fig. 9a, b, c, d. Electromicrographs of four groups of cholera phages. (See pages 83–88.)

phage lysates contained phage neutralising antibodies. Otto and Winkler (1922), using an adsorption test with the host bacteria, demonstrated that phage is antigenically distinct from its host. Later studies have proved the value of antiphage sera in bacteriophage research. Lanni and Lanni (1956) demonstrated that T2 phage particles contain at least two distinct antigens. One antigen is localised on the phage tail and reacts with phage neutralising antibody, while another (precipitating and complement-fixing antigen) is

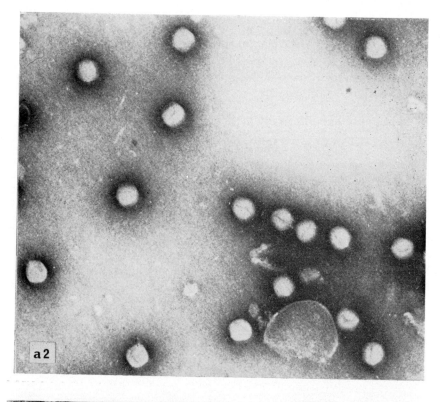

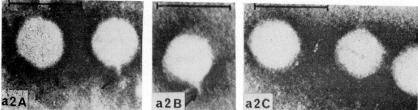

FIG. 9a (*cont.*)

situated on the head. The internal contents of phage obtained by osmotic shock are serologically inactive (Hershey and Chase, 1952; Hershey, 1955).

Antigenic specificity constitutes the most important criterion in the classification of bacteriophages acting on a particular bacterium. Serologically unrelated phages are grouped into separate types. In the case of different types of phages which overlap serologically, the rates of neutralisation as expressed by their K values also differ. The rate of neutralisation by an antiphage serum is greatest with the homologous type. The host range mutants of a phage are serologically identical (Craigie and Yen, 1938; Luria, 1945).

Asheshov *et al.* (1933) studied the antigenic specificity of A, B and C types of *V. cholerae* phages and found them to be serologically distinct. But Pasricha *et al.* (1936) had found that cholera phage A was immunologically distinctive, whereas the other two phages were serologically related.

The *V. cholerae* typing bacteriophages were classified in the beginning into four groups by their lytic patterns on test strains of *V. cholerae*. The classification was confirmed by other tests including antigenicity. Antiserum neutralisation tests (Mukerjee, 1962) were carried out by a method

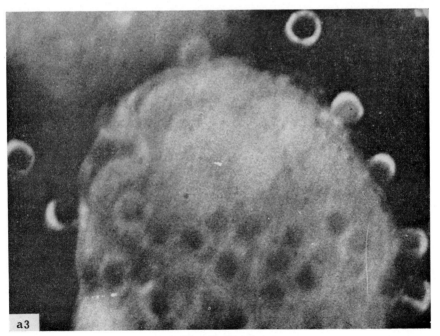

Fig. 9a (*cont.*)

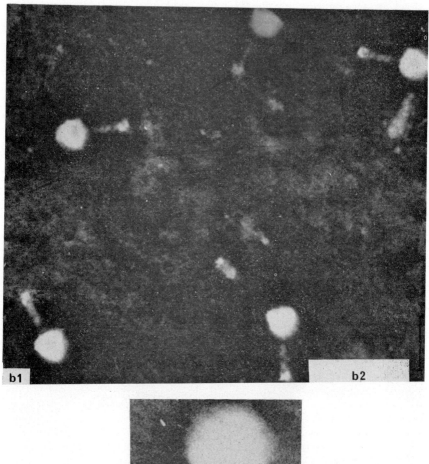

Fig. 9b.

of inhibition of the homologous and heterologous phages after 30 minutes incubation of the phage-serum mixture. The results are given in Table VI.

All the antiphage sera could neutralise the corresponding phages. On further tests it was found that each of the sera inhibited plaque formation by phage strains belonging to the same group roughly to an equal extent. The rate of inhibition of homologous phages by their antisera was tested after 5 and 10 minutes incubation of the phage-serum mixture (Adams, 1959). The results are given in Table VII.

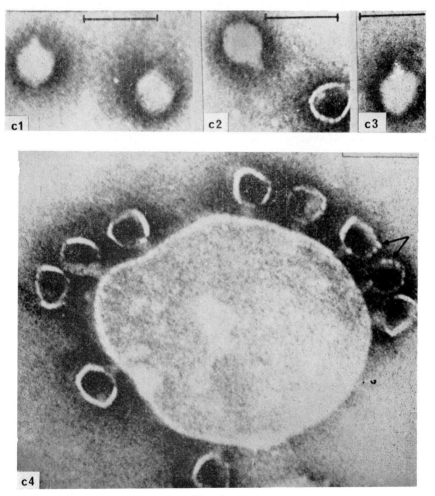

FIG. 9c.

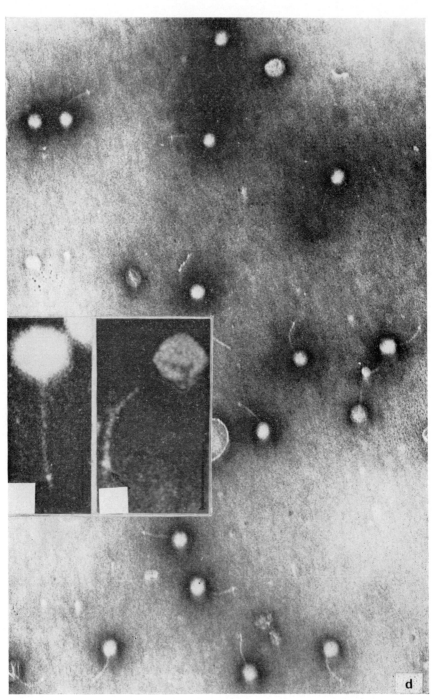

Fig. 9d.

## TABLE VI

### Titration of antiphage sera with homologous phages

| Antiserum against phage group | Group of phage tested | Plaques obtained after reaction with serum dilution of | | | | Control plaques formed by the corresponding dilution of phage |
|---|---|---|---|---|---|---|
| | | 100 | 200 | 400 | 800 | |
| I | I | Nil | Nil | Nil | 23 | Scl |
| II | II | 2 | 12 | 180 | Plq++ | Scl |
| III | III | 140 | 410 | Plq++ | Plq+++ | Scl |
| IV | IV | 2 | 42 | Plq+ | Plq+++ | Scl |

Plq +, + + and + + + indicates plaques from a few to a fair number; Scl, Semiconfluent lysis.

S. MUKERJEE

## TABLE VII

### Inhibition of phages by specific antisera

| Antiphage serum for phage group | Group of phage used in the test | Percentage of inhibition on incubation for | |
|:---:|:---:|:---:|:---:|
| | | 5 min | 10 min |
| I | I | 93 | 97 |
| II | II | 88 | 93 |
| III | III | 68 | 75 |
| IV | IV | 91 | 99 |

Antiphage sera were also used to test antigenic relationships between the groups of typing phages. The results are given in Table VIII which represents the averaged results of four experiments.

Significant inhibition was caused by reaction between phage and its homologous antiserum only; this suggested that the four groups of phages were antigenically unrelated.

## TABLE VIII

### Inhibition of phages by specific antisera

| Antiserum against phage group | Phage group tested | No. of plaques after incubation for 10 min with antiserum | No. of plaques in control without serum | Inhibition |
|:---:|:---:|:---:|:---:|:---:|
| I | I | 1 | 28 | + |
| | II | 42 | 37·5 | − |
| | III | 27 | 28 | − |
| | IV | 20 | 23·5 | − |
| II | I | 29·5 | 28 | − |
| | II | 1 | 37·5 | + |
| | III | 26·8 | 28 | − |
| | IV | 27 | 23·5 | − |
| III | I | 20 | 28 | − |
| | II | 45 | 37·5 | − |
| | III | 8 | 28 | + |
| | IV | 22·5 | 23·5 | − |
| IV | I | 25 | 28 | − |
| | II | 37 | 37·5 | − |
| | III | 25·5 | 28 | − |
| | IV | 0·5 | 23·5 | + |

## D. Host range of cholera bacteriophages on *V. cholerae* (classical)

Thirty cholera phages isolated from stool samples in Calcutta were tested with 200 *V. cholerae* (classical) strains isolated in the same epidemic between May 1955 and February 1957. On the basis of lytic patterns, the 30 phages could be differentiated into four distinct groups (Mukerjee *et al.*, 1957) (Table IX).

TABLE IX

**Sensitivity of 200 *V. cholerae* strains to cholera phages**

|  | Cholera bacteriophages | *V. cholerae* | |
| --- | --- | --- | --- |
| Groups | No. of bacteriophages belonging to the group | No. of strains lysable | No. of strains not lysed |
| I | 23 | 185 | 15 |
| II | 3 | 178 | 22 |
| III | 3 | 197 | 3 |
| IV | 1 | 200 | .. |

The 200 *V. cholerae* strains tested in this series were classified into five phage types on the basis of their sensitivity patterns to four groups of typing phages as given in the Table X.

TABLE X

**Patterns of sensitivity of 5 phage types of *V. cholerae***

| Phage type | No. of strains | Sensitivity to phage group | | | |
| --- | --- | --- | --- | --- | --- |
|  |  | I | II | III | IV |
| 1 | 174 (87%) | + | + | + | + |
| 2 | 4 (2%) | − | + | + | + |
| 3 | 11 (5·5%) | + | − | + | + |
| 4 | 8 (4%) | − | − | + | + |
| 5 | 3 (1·5%) | + | + | − | + |

+ = Sensitive.    − = Resistant.

## E. Host range of cholera bacteriophages, other than the classical type *V. cholerae*

Bacteriophages vary in their lytic specificity for the host bacterium. While some bacteriophages are lytic for single species of bacteria, others possess multiple virulence and affect bacteria belonging to different genera (D'Herelle, 1926). In the first group, the lytic ranges of some are restricted to only a small number of strains. At the species level, the host range pattern forms a valuable characteristic for strain identification. Stocker (1955) suggested that host-range studies may prove taxonomically useful for *V. cholerae*. It had been observed by a number of workers including Burnet (1927), Burnet and McKie (1930), and Craigie and Yen (1938) that a high correlation existed between the heat-stable somatic bacterial antigen and bacterial susceptibility to a given phage. Loss or acquisition of antigens by bacteria resulted in resistance or susceptibility to phage. Burnet (1930) further observed that host susceptibility depended on certain host-cell antigens which did not necessarily act as agglutinogens. Sertic and Boulgakov (1935) reported on a correlation between flagellar antigen and phage susceptibility.

*V. cholerae* biotype *eltor* possesses a common somatic as well as flagellar antigens of classical *V. cholerae* (Abdoosh, 1932). NAG vibrios are not agglutinated with cholera O sera. However, some of the latter possess common H antigen with true cholera vibrios (Taylor *et al.*, 1937; Ahuja and Singh, 1939; Sakazaki *et al.*, 1963; Bhattacharyya and Mukerjee, 1974). Moreover, the rough and rugose antigens of *V. cholerae* and NAG vibrios may have common components (White, 1940). White has further noted that the heat-labile somatic protein (HLSP) antigens of some of the NAG vibrios were precipitated by anti-cholera sera and were different from the ag-glutinogens. Some *Escherichia coli* strains are known to share O antigen with *V. cholerae* (Felsenfeld *et al.*, 1951; Malizia, 1954).

Phage susceptibility patterns of *V. cholerae* strains have been reported (Mukerjee *et al.*, 1957, 1959, 1960). Investigations on the host-range of the four recently isolated groups of cholera bacteriophages in respect of *V. cholerae* biotype *eltor* and NAG vibrios and *E. coli* show that:

1. All the NAG vibrios were insensitive to the group II cholera phage.
2. Group IV phages were totally inactive against any of the *V. cholerae* biotype *eltor* strains while they were universally lytic for *V. cholerae* (classical biotype) which were serologically and biochemically identical with the former biotype.
3. None of the *E. coli* strains tested was susceptible to any cholera phage.

It was inferred that although antigenic structures of bacteria play some role in their susceptibility to bacteriophages, other factors in the surface mosaic appear to determine their susceptibility.

## F. Thermal death points of *V. cholerae* typing phages

Thermal inactivation of bacteriophages is assumed to result from protein denaturation (Adams, 1959) and takes place in accordance with first order reaction kinetics. Working with T5 *E. coli* phages, Lark and Adams (1953) found that phage inactivation by heat is accomplished by liberation of nucleic acid into the solution, the empty ghost phage capsids failing thereafter to become adsorbed on the bacterial surface.

Bacteriophages have shown wide variations in heat resistance, apparently somewhat dependent upon the medium used for propagation and suspension during heat treatment (Burnet, 1930; Gratia, 1940).

Since different batches of non-synthetic medium may vary to some extent, the results of heat susceptibility tests may not be precisely reproducible unless one works with a synthetic medium.

Heat resistant genotypic and phenotypic variants occur in wild-type phage stocks at different frequencies and heat inactivation rates of all the phage particles in a stock may therefore not be uniform. Lark and Adams (1953) have shown the frequency of mutants of T5 phage to heat resistance to be about 1000 times greater than that of the parent wild-type in 0·1 N NaCl.

Thermal death point, like the serology, morphology and host range patterns, is one of the parameters for identification of bacteriophages. The heat susceptibility of the four groups of *V. cholerae* (classical) typing phages has been studied (Mukerjee, 1961b). The results are given in Table XI.

In any phage stock, a few PFU occasionally survived exposure to a temperature at which the vast majority of the remaining particles were inactivated. In our studies, attempts were made to purify *eltor* biotype typing phages by repeated pure-line isolations. The tenth pure-line phage isolate showed wider variations in thermal death point than the original stock. This was not due to host-mediated transformation, as all the phages were propagated on the same propagating strain of *V. cholerae* biotype *eltor* No. Mak. 757.

## G. Thermal and pH inactivation of biotype *eltor* phage Ph-1

The *eltor* biotype phage Ph-1 is large (1400–2300 Å) and has a loosely packed structure. Maiti and Chatterjee (1969a) studied its physicochemical characteristics in pH and thermal inactivation tests.

At pH values in nutrient broth ranging from 5·0 to 10·0 for an incubation at 37°C for 1 hour this phage appeared stable. Below pH 5·0, there was a very sharp decline in survival of the phage and from pH 4·0 downwards the phage was completely inactivated. In spite of its unusually large size compared to other bacteriophages, its pH susceptibility was as observed for other phages.

## TABLE XI

### Inactivation of cholera phages after exposure to different temperatures for 30 min†

| Phage No. | Phage group | Survival of phage after exposure (temperature °C) | | | | | | | | Range of thermal death points (°C) |
| --- | --- | --- | --- | --- | --- | --- | --- | --- | --- | --- |
| | | 58 | 60 | 63 | 65 | 68 | 70 | 73 | 75 | |
| 191 | I | Cl | Cl | Cl | Scl | Inn | 1 Plq | — | — | 65–70 |
| 163 | I | Cl | Cl | Cl | Scl | Inn | — | — | — | 65–70 |
| 138 | II | Cl | Cl | Cl | Scl | Scl | Scl | — | — | 70–73 |
| 186 | II | <Cl | <Cl | <Cl | <Cl | <Cl | Scl | — | — | 70–73 |
| 145 | III | Inn | 100 Plqs | 3 Plqs | — | — | — | — | — | 60–63 |
| 185 | III | Cl | Cl | 1 Plq | — | — | — | — | — | 60–63 |
| 149 | IV | Cl | Cl | Cl | Cl | Inn | 110 | — | — | 68–73 |
| MCIV | IV | Cl | Cl | Cl | Cl | Inn | 500 Plqs | — | — | 68–73 |

†Cl = Confluent lysis, <Cl = Less than confluent lysis, Scl = Semi-confluent lysis, Inn = Innumerable plaques, Plqs = Plaques

## H. Phage-resistant mutants of *V. cholerae*

Susceptible bacteria may develop resistance to bacteriophages as a result of mutation or lysogeny.

Isolation of phage-resistant mutants from a culture is performed by plating the bacterial culture and an excess of phage when the resistant mutants may form colonies.

Asheshov *et al.* (1933) found that mutants of vibrios resistant to one type of phage could still be susceptible to the action of other types. The phenomenon was investigated in detail and reciprocal cross reaction of bacteriophages with phage-resistant secondary growths formed the principal basis of classification of all types of cholera bacteriophages isolated. The *V. cholerae* (classical) typing bacteriophages were classified into groups primarily on the basis of lytic patterns for *V. cholerae* (Mukerjee *et al.*, 1957). Mutant strains of *V. cholerae* resistant to these phage groups were prepared. Their phage sensitivity patterns (Mukerjee, 1962) are given in Table XII and Fig. 10.

### TABLE XII
**Single mutational pattern of a type 1 *V. cholerae* strain**

| Group of phage used for mutant selection | Pattern of sensitivity of the mutant | | |
|---|---|---|---|
| | Resistant to phage group | Sensitive to phage group | S-R dissociation |
| Nil | Nil | I, II, III & IV | S |
| I | I & II | III & IV | R |
| II | II | I, III & IV | S-R |
| III | III | I, II & IV | S |
| IV | IV | I, II & III | S |

S = Smooth, S-R = Smooth-rough, R = Rough.

## I. Neutralisation of cholera bacteriophages by extracts of *V. cholerae*

Inactivation of bacteriophages by bacterial extracts was first demonstrated by Levine and Frisch (1934). Later, Burnet (1934) suggested that (i) heat-stable surface antigens are specially correlated to phage-susceptibility of the host bacteria and (ii) carbohydrate-haptens of the cell surface serve as virus receptor sites. The bacterial component inhibiting the phages was termed "phage-inhibiting agent (PIA)". Some of the properties and the mode of action of PIA *vis-à-vis* antiphage sera were demonstrated. On the basis of his experiments, Burnet suggested that the sites of action of

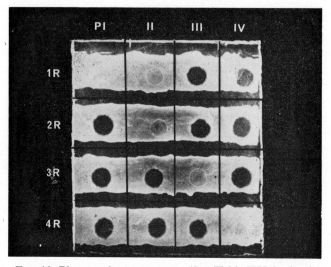

FIG. 10. Phage resistant mutants. (See Table XII for key.)

PIA and phage antibodies were adjacent and not identical. Miller and Goebel (1949) working with phage types I and II *Shigella sonnei* on T phages, concluded that the receptor sites for the two groups of phages were different. The possible chemical nature of PIA was indicated by various other workers (Goebel and Perlman, 1949; Goebel, 1950; Jesaitis and Goebel, 1952; Miller and Goebel, 1949; Goebel and Jesaitis, 1953) while working with purified antigen of *Sh. sonnei* on T phages.

Differences in the inhibiting action of different extracts of *V. cholerae* strains (including El Tor and NAG vibrios) on four groups of phages used for typing *V. cholerae* (Mukerjee *et al.*, 1957) were studied by Ghosh and Mukerjee (1960).

Groups I and IV phages were not inhibited by any of the antigenic extracts, whereas groups II and III phages were inhibited by the extracts of some of the vibrio strains. The extracts of all the five phage types of *V. cholerae* and El Tor strains completely inhibited group II phages irrespective of their susceptibility to this phage group. But phage inhibiting agent (PIA) for group II phages was present in higher concentration in the extracts of type 1 and other sensitive strains than in those of strains resistant to it. None of the extracts of NAG strains was capable of inhibiting group II phages.

Group III phages were inhibited only by extracts of those strains which were sensitive to this phage group. Extracts treated with acetic acid showed less phage inhibiting activity. But when the extract was fractionated into soluble and insoluble portions by addition of 0·05 M TCA, phage inhibiting activity was detectable in both the fractions. No attempt was made to determine the precise chemical nature of the phage-inhibiting agent.

## J. A phage specific for the pathogenic strains of *V. cholerae* biotype *eltor*

A comparative study (Basu and Mukerjee, 1970) of strains of *V. cholerae* biotype *eltor* from different sources in respect of their pathogenicity and phage sensitivity is given in Table XIII.

TABLE XIII

**Human and animal pathogenicity of biotype *eltor* strains of *V. cholerae* and their sensitivity to the group V typing phage Ø (H74/64)**

| No. of strains | Source of isolation | Animal pathogenicity | | Phage sensitivity‖ |
| --- | --- | --- | --- | --- |
| | | Rabbit ileal loop test | Infant rabbit test | |
| 3464 | Cholera El Tor patients | Positive† | Not done | Sensitive |
| 4‡ | Cholera El Tor patients | Negative | Negative | Resistant |
| 32§ | Water sources | Negative | Negative | Resistant |
| 600 | Classical cholera | Positive | Positive | Resistant |

† Only 60 representative strains tested.

‡ These 4 strains were received from Mrs C. Z. Gomez, Department of Health of the Philippines, Manila. Pesigan *et al.* (1967) classified them as "atypical" after applying all the commonly employed tests for characterisation of El Tor vibrios.

§ Isolated from Middle East countries and water sources in India, Burma etc. before the current pandemic; these were O-I positive.

‖ At 1 RTD and 100 RTD.

It may be seen that most of the *V. cholerae* biotype *eltor* isolated from cholera patients and giving a positive rabbit loop reaction were sensitive to phage H74/64, while the 32 strains isolated from water neither gave a positive ileal loop reaction nor were sensitive to phage; this showed that human and animal pathogenicity of these strains ran closely parallel. The four "atypical" strains presented by Pesigan *et al.* (1967) proved exceptional. These strains differed markedly from those isolated from cholera patients since 1937, were strongly haemolytic, but differed from the biotype *eltor* strains in being non-haemagglutinating with biochemical reactions of Heiberg Group III. They were sensitive to Polymyxin B, but insensitive to group IV phage. They were non-lysogenic, but bacterio-cinogenic. Lack of phage sensitivity of these strains does not diminish the value of the results obtained with a large number of other strains of this biotype. Use of phage Ø H74/64 at RTD may therefore be helpful in differentiating pathogenic El Tor strains from the classical ones, NAG vibrios, and also non-pathogenic variants of the biotype *eltor*.

## K. Kappa-type phages

Takeya and Shimodori (1963) of Kyushu University, Japan, reported that *V. cholerae* biotype *eltor* isolated from epidemics on the West Pacific Islands and South East Asia were lysogenic with a special type of temperate phage having a very narrow host range and designated by them as "Kappa-type" phage. Chun *et al.* (1970) confirmed the constant association of Kappa-phage with strains isolated in cholera El Tor epidemics in South Korea. Neogi and Sanyal (1966) also found the same type of phage in *eltor* biotype strains isolated in the Calcutta outbreaks. On the other hand, Takeya *et al.* (1967) found that some stock cultures of strains of *V. cholerae* biotype *eltor* were non-lysogenic, but resistant to lysis by Kappa-type phages. It was possible to lysogenise these strains with Kappa phages. He, therefore, designated them as "cured" strains of Celebes type as compared with the classic, Ubon, type which were non-lysogenic and susceptible to lysis by the Kappa phages. In fact, he found up to 3% of El Tor strains isolated in epidemics to be of the "cured" type and all strains isolated in the epidemics of 1966 in Thailand and Cambodia were found free from lysogenic phages.

In 1963, when Takeya and Shimodori first published their paper on Kappa-phage, they suggested that there might be a causal relationship between lysogeny and virulence in El Tor vibrio similar to lysogeny and toxigenicity in *Corynebacterium diphtheriae*. However, later observations did not confirm their hypothesis.

Kappa-type phages have been studied in detail by Takeya *et al.* (1965a, b). According to Takeya (1974), these phages have a very narrow host range. A classical type strain H/218 is used as the indicator organism and only a few other strains of *V. cholerae* are susceptible to them. Plaques of all Kappa-phages on H/218 strain are turbid with occasional clear plaque mutants. The host ranges of these mutants are usually broader than those of the parent phage. All Kappa-phages, including their mutants, are antigenically identical and they are morphologically identical (Fig. 11). According to Takeya (1974), the head is hexagonal, 450 to 550 Å in diameter. The length of the tail is 800 to 1000 Å. Its width is 150 Å. Concentration of Kappa-phage in broth culture reaches a maximum of $10^{3-6}$/ml after 24-28 hours incubation. Exposure to ultra-violet irradiation and Mitomycin C (Takeya,

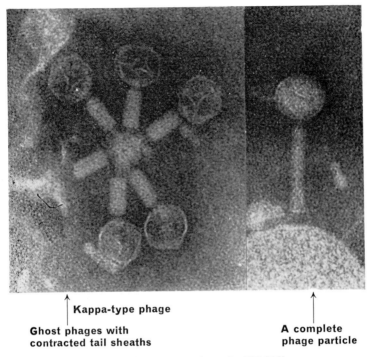

**Kappa-type phage**

**Ghost phages with
contracted tail sheaths**

**A complete
phage particle**

FIG. 11. Kappa-type phage ( × 260,000).

1967) can induce Kappa-phage production. Kappa-phages are more sensitive to heat, but more resistant to UV as compared with the *V. cholerae* typing phages.

To detect Kappa-phages in a stool specimen, it is added to alkaline peptone water; after incubation for 12–18 hours a spot test is performed on H/218 Streptomycin-resistant indicator strain plated on agar media containing Streptomycin. Following incubation at 37°C for 8 hours the presence of lysis indicates Kappa-type phages in the sample.

## L. Characteristics of typing phages for *V. cholerae* biotype *eltor*

All the five bacteriophages used by Basu and Mukerjee (1968) for phage typing of *V. cholerae* biotype *eltor* strains were obtained either from stools or from lysogenic *V. cholerae* strains. However, there remained a probability that these phage preparations might not be quite pure. Therefore, pure line isolations were carried out repeatedly by Dutta *et al.* (1972). On plating Group II phages, two plaque types were obtained – one slightly bigger than the other. Based on this slight difference in plaque morphology, both the plaque types were picked separately and cloned eleven times. For the time being, these two types were designated as Group IIA and Group IIB phages. The other four phages gave plaques of uniform morphology. The biological characteristics of these 6 phages, viz. plaque morphology, thermal death points, antigenic composition, dynamics of generation and patterns of sensitivity to these phages among phage resistant mutants were studied. The results are given in Tables XIV, XV, XVI and XVII.

It may be seen from Table XVII that the thermal death points of pure line phages were not the same under identical conditions of experiments. The thermal death points of each phage also varied with the length of exposure. Group IIB phage proved more resistant to heat than Group IIA phage. Group III and Group IIB phages were most heat resistant while Group I was least resistant to heat.

Table XV illustrates the antigenic relationship between the 6 pure line phages. It may be seen that antiphage sera against Groups I, IIA and IIB neutralised Group III phage. Antiphage serum against Group IIB also neutralised Group V. Antiserum against Group IV did not neutralise any of the heterologous phage Groups and serum against Group V phage also

TABLE XIV

**Thermal death points of six pure line phages**

| Phage group | 20 min | 30 min |
|:-----------:|:------:|:------:|
| I | $63° - 70°C$ | $63° - 68°C$ |
| IIA | $68° - 73°C$ | $65° - 70°C$ |
| IIB | $68° - 75°C$ | $68° - 73°C$ |
| III | $68° - 75°C$ | $68° - 73°C$ |
| IV | $68° - 73°C$ | $65° - 70°C$ |
| V | $68° - 73°C$ | $65° - 70°C$ |

TABLE XV

**Antigenic relationship between six pure line phages**

| Antiphage serum against phage groups | Phage group tested | | | | | |
|:---:|:---:|:---:|:---:|:---:|:---:|:---:|
| | I | IIA | IIB | III | IV | V |
| I | + | − | − | + | − | − |
| IIA | − | + | − | + | − | + |
| IIB | − | − | + | + | − | − |
| III | − | − | + | + | − | − |
| IV | − | − | − | − | + | − |
| V | − | − | + | − | − | + |

## TABLE XVI

**Patterns of sensitivity of single-step phage resistant mutants of type 1 *V. cholerae* biotype *eltor***

| Groups of phage for mutant selection | Code number of the resistant strain | Lytic action with phage group | | | | | | S-R dis-sociation |
|---|---|---|---|---|---|---|---|---|
| | | I | IIA | IIB | III | IV | V | |
| Nil | Mak 757 | + | + | + | + | + | + | S |
| I | IR | − | + | + | + | + | + | SR |
| IIA | IIAR | + | − | + | − | − | + | SR |
| IIB | IIBR | + | − | − | − | + | − | R |
| III | IIIR | + | − | − | − | − | − | R |
| IV | IVR | + | − | + | + | − | + | SR |
| V | VR | − | + | + | − | + | − | SR |

+ = Sensitive; − = Resistant; S = Smooth; SR = Smooth rough; R = Rough

## TABLE XVII

**Dynamics of generation of pure-line phages**

| Phage group | Time of appearance of plaques | Latent period of generation |
|---|---|---|
| I | $2\frac{1}{2}$ – 3 h | 30 – 35 min |
| IIA | 3 – $3\frac{1}{2}$ h | 35 – 40 min |
| IIB | $2\frac{1}{2}$ – 3 h | 30 – 35 min |
| III | $2\frac{1}{2}$ – 3 h | 30 – 35 min |
| IV | $3\frac{1}{2}$ – 4 h | 45 – 50 min |
| V | $2\frac{1}{2}$ – 3 h | 30 – 35 min |

inhibited Group IIB phage. From these results it seems that Group IIB and III phages are closely related, as also are Group IIB and Group V phages; but they do not have identical antigenic constitution, because the antiphage serum prepared against Group I and Group IIA phages neutralised Group II phages, but had no activity against Group IIB and Group V phages. Again, Group III antiphage serum neutralised Group IIB phage. Unlike classical cholera phages (Mukerjee, 1962), all phage groups except the Group IV phage were antigenically interrelated to varying degrees and each of them had multiple antigenic structures of which only a part is involved in lytic specificity.

Table XVI shows the patterns of sensitivity of single-step resistant mutants of Mak 757. A mutant resistant to Group I phage is also resistant to Group III phages. A mutant resistant to Group IIA phage is also resistant to Group IV phage; that prepared against Group IIB phage showed resistance to phage Groups IIA, III and V. It is therefore evident that unlike the classical *V. cholerae* phages (Mukerjee, 1962), where resistance was strain specific, phage resistant mutants of *V. cholerae* biotype *eltor* appeared resistant both to the particular phage involved and to other phage groups. Multiple resistance of one-step mutants has also been reported by Luria (1945) and Anderson (1946) with *E. coli* strains B and T phages. They suggested that due to overlapping of the receptor sites on the surface of bacterial cells, the failure of synthesis of the receptor substances in mutants could also affect the receptor mechanism for other phages.

## IX. BACTERIOCINOGENY AND BACTERIOCIN TYPING OF *V. CHOLERAE*

Bacteriocinogeny and sensitivity to bacteriocin has been successfully applied as an epidemiological tool in several bacterial species. Farkas-Himsley and Seyfried (1962) first introduced the term vibriocin to describe a lethal substance synthesised by a lysogenic strain of *V. cholerae* which was active on its Streptomycin-resistant mutant. The purified substance was found to be particulate, resembling bacteriophage tails, and although it was neither structurally nor immunologically related to Group IV phage, it shared a common receptor site (Jaywardene and Farkas-Himsley, 1968). An association between the genetic acquisition of vibriocinogeny and resistance to Group IV phage has been demonstrated by Takeya and Shimadori (1969).

Takeya has reviewed the several attempts made to demonstrate vibrio-cinogeny on a routine basis (Mukerjee and Takeya, 1974). Using entero-bacterial colicin indicator strains in the conventional cross-streaking test of Abbott and Shannon (1958), Barua (1963) was unable to demonstrate bacteriocinogeny in 75 strains of *V. cholerae*. By introducing several modifications into the test however, Chakrabarty *et al.* (1970) were able to demonstrate strong inhibitory activity by *V. cholerae* in several common media. The essential factors in their test were the presence of citrate (Chakrabarty and Dastidar, 1974) and treatment of the producer culture by cold shock for 18 hours although Bhaskaran *et al.* (1974) have questioned the specificity of the test. The vibriocin of Chakrabarty and his co-workers would appear to be different from that of Farkas-Himsley and Seyfried (1963) in being unstable and heat sensitive. On the basis of their results with 425 O group I *V. cholerae* and 215 strains of other serotypes, Chakrabarty and his co-workers (1970; 1971a; 1971b) were able to distinguish a total of 15 types. However, no correlation between bacteriocin type, biotype, serotype, phage type or epidemiological history could be demonstrated in this series of strains (Chakrabarty *et al.*, 1970). A bacteriocin typing scheme of practical value remains to be developed.

## X. TECHNIQUES USED IN CHOLERA BACTERIOPHAGE STUDIES

### A. Double agar layer method of plating a phage

For phage typing, 0·1 ml of a suitable dilution of the phage preparation and 0·1 ml of a 4 hour nutrient broth culture of its propagating strain are added to sterilised tubes of molten nutrient agar (0·7%), kept at 42°C and the contents mixed well. The mixture is then poured onto a dry nutrient agar plate. Mixing and even distribution of phage and bacteria are facilitated by rocking the plate to and fro for a few seconds. The plates are left on the bench for about 15 minutes to allow setting. They are dried briefly at 37°C and then incubated overnight. Plaques appear as clear areas within the translucent lawn of growth of the host propagating strain.

### B. Phage propagation in liquid media

A few drops of a suspension of the phage to be propagated and 0·5 ml of a young broth culture of *V. cholerae* (strain 154 for all classical strains) are added to 50 ml of nutrient broth in a 250 ml conical flask to provide for a large surface and shallow layer. *V. cholerae* Makassar 757 is employed similarly for propagation of El Tor typing phages. The propagating strains for NAG vibrio phages are different and several are in use.

### C. Phage propagation in solid media

A mixture of 0·1 ml of phage and a young broth culture of the host propagating strain is spread over a nutrient agar plate, the excess fluid being removed with a Pasteur pipette, the surface moisture removed by drying, and the plate incubated overnight at 37°C and read thereafter. Alternatively, 0·1 ml of the phage preparation in serial ten-fold dilutions and 0·1 ml of a 4 hour culture of the propagating vibrio are mixed with soft agar and layered over a nutrient agar plate and incubated overnight. The highest dilution of the phage lysate which gives confluent lysis is selected. The soft agar is scraped off the plates and mixed well with 15–20 ml nutrient broth and centrifuged to obtain a clear supernatant. Subsequent treatments are as described above. The phage preparations obtained from the agar surface usually contain $10^{12}$ to $10^{14}$ plaque forming units (pfu) per ml. This, when added to liquid propagation medium, yields similar titres.

## D. Preparation of high titre cholera phages

The concentration of phage preparations ($10^{12}$ pfu per ml) depends on the density of growth of bacteria in the logarithmic phase and addition of appropriate quantities of phage. Heavy growth of the host propagating strain in liquid culture can be obtained by using e.g. 3% peptone or papain digest broth and aeration of the culture either by shaking (in a mechanical shaker) or by bubbling air or oxygen. Phage is added to logarithmic phase cells in multiples of 5 to 10 to the number of host propagating bacteria to ensure phage infection of all cells. After a specified incubation period, the unaffected cells and debris are spun down by low speed centrifugation to minimise loss of phage by adsorption. Loss of phage due to filtration may also be avoided by treatment with heat or chemicals which kill bacteria, but leave the phage unaffected. Most cholera bacteriophages withstand a temperature of 55°C for half an hour, toluene or chloroform, while the propagating strain is killed by such treatments. Recently, Monsur *et al.* (1970) reported the preparation of a very high titre phage (of an order higher than $10^{12}$ pfu per ml) by the technique of high speed centrifugation and suspension of the phage pellet in one tenth the original volume. The resultant phage preparation was used by them for evaluating the effect of massive doses of bacteriophages for cholera therapy.

## E. Sensitivity tests of *V. cholerae* to vibrio phages

The principle of phage susceptibility tests of vibrio strains is similar to that of Craigie and Yen (1938) for phage typing of *Salmonella typhi*. A lawn, spot or streak culture of the test *V. cholerae* strain is made on the surface of a nutrient agar plate, standard volumes of the typing phages being spotted at marked sites on the plate with the help of standardised loops or droppers. Plates are read after overnight incubation at 37°C.

## F. Determination of routine test dilution (RTD) of a phage

To determine RTD, ten-fold dilutions of the phage in broth are spotted onto an agar surface just inoculated with a suitable dilution of a broth culture of a sensitive strain of vibrio. The highest dilution producing confluent lysis is known as the critical test dilution or RTD of the phage.

## G. Designation of *V. cholerae* lytic patterns

The technique is designed to obtain areas of uniform inoculation of strains of bacteria under test, on which standard volumes of typing phages at RTD are plated. Readings are taken after overnight incubation.

The different degrees of reaction are recorded as follows:

(i)  Clear or confluent lysis – CL
(ii)  Less than confluent lysis with a few discrete resistant colonies – < CL
(iii)  Semi-confluent lysis – SCL
(iv)  Different degrees of plaque formation – Plq +, + +, + + +
(v)  Opaque lysis with evidence of growth inhibition – OL
(vi)  No lysis – –
(vii)  CLC, < CLC and SCLC are recorded when there are thin layers of growth covering the various degrees of lysis.

## H. Preservation and storage of *V. cholerae* typing phages

Concentrated cholera bacteriophages keep well indefinitely in sealed ampoules at 4°C (Mukerjee *et al.*, 1957). However, diluted phage, e.g. at RTD, loses titre even in the refrigerator. Bacteriophages can also be preserved (especially the Group IV phages) in the dried state on filter paper strips and dispatched in sealed cellophane envelopes if necessary. Phages may also be preserved in the lyophilised state after being suspended in broth or peptone water or any other preservation fluid. Not much work has, however, been done with cholera phages in this direction. For use in laboratories, fresh phage preparations should be derived from stocks preserved either in ampoules or on paper strips. Growth of contaminating bacteria in phage filtrates can be prevented by the addition of a few drops of chloroform.

## I. Phage adaptation

Undiluted high titre phage ($10^{12}$ pfu per ml) and serial ten-fold dilutions are plated by the agar layer method on an insensitive bacterial strain. If plaques are found, they are picked directly. In their absence, 5 ml of the high titre phage (undiluted) is added to 1 ml of a 3 to 4 hour broth culture of the strain and the mixture incubated overnight at 37°C. The procedure for phage propagation is followed and the filtrate tested for presence of plaque forming units by the agar layer method. In case of failure, the process is repeated up to 6 times. Plaques so obtained are propagated separately, purified, and their host ranges determined. If the original host range is obtained after propagation with the original propagating strain, it is considered to be an *adaptation;* if not, it is considered a host range *mutation.*

Phage adaptation has been successfully used by Mukerjee *et al.* (1957) for subtyping type 2 strains of classical *V. cholerae* and by Dutta *et al.* (1972) for the development of an improved phage typing scheme for *V. cholerae* biotype *eltor* strains. The same technique was also used by Sil *et al.* (1974) for the development of a bacteriophage typing scheme for *V. cholerae* strains other than O serotype I.

## J. Induction of lysogenisation of *V. cholerae*

A virulent phage infects, replicates vegetatively and lyses susceptible bacteria, while a temperate phage enters into a symbiotic relationship with the susceptible bacteria and exists as a prophage in the bacterial gene replicating along with the latter; here the bacteria remain alive and become lysogenic. Occasionally the prophage reverts to vegetative phage at a very low frequency. With the use of "inducing" agents like ultraviolet light, X-rays, or chemical inducers, the production of lytic mutants is accelerated. The prophage in lysogenic bacteria is non-infective, but once lysogenic phages are liberated they are capable of lysing or lysogenising fresh susceptible bacteria (Adams, 1959). The relative frequency with which these two responses occur depends on the conditions of infection and the genetic constitution of the phage and the host.

In the process of lysogenisation of sensitive bacteria, some temperate phages are able to carry a piece of chromosomal material from the donor bacteria to the recipient conferring new hereditary characters on the latter (Zinder and Lederberg, 1952).

## K. Preparation of antiphage serum

Antiphage serum is prepared by immunising rabbits weighing 1·3 to 1·5 kg. Five ml of a high titre phage is injected twice weekly. The first two injections are given subcutaneously, subsequent ones intravenously. Test bleeding is performed one week after the last injection. If satisfactory antibody titres are not obtained, a further course of 3 injections is given. Blood is collected aseptically. The serum after separation is kept in the refrigerator in sealed ampoules without preservatives.

A preliminary bleeding before immunisation of the rabbit should be carried out in every case. In phage neutralisation tests, the dilutions of the pre-immunisation serum used are the same as those of the immune serum from the same rabbit in control and test respectively, to exclude non-specific neutralisation of the phages.

## L. Titration of antiphage serum

Antiserum neutralisation tests are performed by a method similar to those of Adams (1959) and Anderson and Felix (1953). For preliminary assay of antiphage serum, the bacteriophage stock is diluted suitably in nutrient broth (pH 7·4) to produce semiconfluent lysis or innumerable discrete plaques when 0·05 ml is plated by the agar layer method. The antiphage serum is also diluted in two-fold dilutions starting from 1/50 in the same medium. Equal parts of phage dilutions and progressive serum dilutions are mixed in separate tubes and put into a water-bath at 37°C for half an hour. The mixture (0·1 ml) is then plated by the agar layer method on each of a set of three plates with a young culture of a lysable strain of *V. cholerae*. The inhibition in plaque count in the test series as compared with a control set indicates the neutralising titre of the serum under the conditions of the test.

For determining the antigenic relationship between the groups of cholera phages, phage stocks are diluted to a titre of about $10^7$ pfu per ml and the antiphage serum to 1 in 50 in broth. One volume of each of the diluted phage groups is mixed with 99 volumes of the serum dilution and put into a water-bath at 37°C for 10 minutes for neutralisation. Aliquots of the mixture from all the tubes are then diluted 1 in 100 in broth to stop further action of the serum. One standard loopful delivering 0·01 ml of these dilutions is then plated on each of four plates by the soft agar layer method. If no inactivation by serum occurs, 25–50 plaques will appear after incubation.

In cross-neutralisation tests to determine antigenic relationships, experiments should be adequately controlled by including the corresponding serum from the pre-immunisation bleeding.

## M. Method of detection of Kappa-phage for tracing carriers of *V. cholerae* biotype *eltor*

To detect Kappa-type phage in the stools of *V. cholerae* biotype *eltor* carriers, it is necessary to enrich the phage by addition of the indicator strain to the peptone water culture of the stool. After 12–18 hours incubation, a spot test is performed on a lawn of the Streptomycin resistant indicator strain H.128 on a Streptomycin containing agar plate. Evidence of lysis indicates the probable presence of Kappa-type phage. Further confirmation may be obtained by using a Kappa-phage resistant H.218

*Scheme of phage detection method for routine use*

Specimens
↓
Enrichment culture in alkaline peptone water for 12–18 hours

Isolation of *V. cholerae* by ordinary cultural methods

Spot test on H.218 Sm$^r$ seeded in Streptomycin containing agar plate
    Incubate at 37°C
    for 8–12 hours
↓
Lysis ( + ) - probable presence of Celebes type of *V. cholerae* biotype *eltor*

If necessary further confirmation by using Kappa-phage resistant H.218 or antiKappa-phage serum.

strain or antiKappa-phage serum. A scheme of the method in use as reported by Takeya (1974) is given here.

## XI. CONCLUDING REMARKS

The aim of all typing systems is to identify strains within a species (or its further subdivisions) with a degree of precision that makes it safe to assume that all the isolates from an epidemic are truly identical and have therefore originated from a single parent strain. It might seem that true identity of two or more strains can be established by examining a large number of biological characters (biotype determination). In theory, even if one chooses to examine a large number of characters, one may not find sufficiently distinctive characters. In practice, for epidemiological surveillance, an elaborate study of each of the isolates may simply not be practicable, even if it were useful.

Of the three methods of typing *V. cholerae:* serotyping, phage-typing and bacteriocin biotyping, serological typing was the earliest to be developed. It has been used in most epidemics and in most laboratories (Pollitzer, 1959) but in a single epidemic in East Pakistan, Bart *et al.* (1970) found the simultaneous incidence of both biotypes and serotypes of *V. cholerae*. In 1968-69 outbreaks, strains of El Tor, Ogawa and classical Inaba *V. cholerae* were isolated from 247 patients and their family contacts. This illustrates the inadequacy of sero- and biotyping of *V. cholerae* strains for epidemiological use. Preliminary biotyping helps in the application of other typing methods such as phage typing. The serotypes of *V. cholerae* have also been found unstable, not only in the course of epidemics, but serotypes have also changed during the course of epidemics. Serotypes have also changed during the course of successive isolations from different epidemics. In fact it has been possible to show changes in Ogawa serotype *in vitro* to the Inaba serotype by cultivation of Ogawa strains in the presence of Ogawa monospecific antisera (Shrivastava and White, 1947); Sack and Miller (1969) reported progressive changes in serotype in germ-free mice infected with *V. cholerae*.

Vibriocin typing is the latest typing method developed for strain identification of *V. cholerae*. It has not yet been possible to study the epidemiological value of the technique for practical use. No standard vibriocin typing method has yet been developed nor is one in routine use. Chakrabarty *et al.* (1960) could not find any correlation between different serotypes, phage types, haemolytic character, years and sources of isolation and bacteriocin types of *V. cholerae*. In fact, a standard bacteriocin typing scheme for practical epidemiological use is yet to be developed.

Phage typing of *V. cholerae* on the other hand has been extensively studied

## TABLE XVIII
### List of sources and reference agencies

| | |
|---|---|
| **1.** *Type Culture Collections* | *Materials available* |
|   (a) WHO Collaborating Centre for Research and Reference on Vibrios, 3 Dr M. Ishaque Road, Calcutta-700016, India. | All type strains, representative strains isolated in different countries in different years, typing antisera and bacteriophages. |
|   (b) American Type Culture Collection, 12301 Parklawn Drive, Rockville, Maryland, U.S.A. | Reference strains of classical *V. cholerae*, *V. cholerae* biotype *eltor*, classical typing phages and classical phage type strains. |
|   (c) National Collection of Type Cultures, Central Public Health Laboratories, Colindale Avenue, London NW9 5HT, U.K. | Reference strains of *V. cholerae* and the serotype strains of Gradner and Venkatraman. |
| **2.** *Individual Scientists* | |
|   (a) Dr R. Sakazaki, Chief, Division of Bacteriology, National Institute of Health, 10–35 Kamiosaki, 2–Chome, Shinagawa ku, Tokyo, Japan. | All type strains of serotypes 1 to 56 of *V. cholerae* (including the so-called NAG vibrios), mucoid strains and R forms. |
|   (b) Professor Kenji Takeya, Department of Bacteriology, School of Medicine, Kyushu University, Fukuoka University, Fukuoka 812, Japan. | Reference strains for Kappa-phage typing. |
|   (c) Dr J. Sil, Cholera Research Centre, Calcutta-700016, India. | NAG bacteriophages. |
|   (d) Dr F. K. Bhattacharyya, Indian Institute of Experimental Medicine, Calcutta-700032, India. | H antiserum. |
|   (e) Dr. S. Basu, Central Research Centre, Kasauli R.I. 173205, H.P., India. | O : I antisera. |

(The international reference preparation of O : I antisera is no longer available.)

| | |
|---|---|
| **3.** *Commercial* | |
|   (a) Difco Laboratories, Detroit, Michigan 48232, U.S.A. | O : I antisera. |
|   (b) Wellcome Reagents Ltd., Langley Court, Beckenham, Kent, BR3 3BS, U.K. | O : I antisera. |

in many outbreaks for over ten years and in a number of laboratories has been found to be of much practical utility. Bacteriophage work requires special laboratory facilities, expertise and care. For successful studies, typing bacteriophages should be maintained in good order and used in high titre. Although this is simple for the classical biotype typing phages, the biotype *eltor* typing phages are comparatively difficult to maintain and to prepare in high titres. Special attention must be paid in maintaining them and also the phage type *eltor* strains, which should be used from time to time to check the proper identity of the phages by their lysis patterns. In case of difficulty the use of antiphage sera can serve very useful purposes for proper identification. It is therefore recommended that combined biotyping and phage typing methods should be used simultaneously for identification of *V. cholerae* for practical epidemiological use. In cases of difficulty isolated *V. cholerae* strains should be referred to any of the *Vibrio* reference laboratories listed in Table XVIII.

## ACKNOWLEDGMENTS

The author is most grateful to Mr. N. K. Dutta for his most willing and painstaking work for preparation of the typescript. I am also indebted to my colleagues, Drs. S. C. Sil, A. Narayanswami, S. N. Chatterjee and F. K. Bhattacharyya for their invaluable help, suggestions and discussions for the improvement of the manuscript. Assistance of the librarian, Royal Society of Medicine, London, and Mr. B. Ghosh of the Indian Institute of Experimental Medicine Library, Calcutta, needs special mention for their valuable help in obtaining some of the rare references.

### Note added in proof

On page 70, it is stressed that O agglutination must be done with fresh cells. Heating cells reduce titres significantly. The editors would like to point out that this is reminiscent of mixed O- and K-antigen agglutination as it is experienced with *Enterobacteriaceae*. Heat should by definition not destroy the lipopolysaccharide O-antigens, whereas envelope antigens may be either washed off or destroyed upon washing or heating the cells. It would seem that further investigations on the chemical nature of the antigens involved are necessary before the agglutination as described is accepted as O-agglutination *sensu strictu*.

### REFERENCES

Abbott, J. D., and Shannon, R. (1958). *J. clin. Path.*, **11**, 71–77.
Abdoosch, Y. B. (1932). *Br. J. exp. Path.*, **13**, 42–51.
Adams, M. H. (1959). "Bacteriophages", Interscience Publishers Inc., New York.
Ahuja, M. L., and Singh, G. (1939). *Ind. J. med. Res.*, **27**, 287–295.

Anderson, E. S., and Felix, A. (1953a). *J. gen. Microbiol.*, **8**, 408.

Anderson, E. S., and Felix, A. (1953b). *Int. Congr. Microbiol.*, *6th Congr.*, *Rome*, *1953*, *Rpt. Proc.* 3, 468.

Anderson, E. S., and Fraser, A. (1956). *J. gen. Microbiol.*, **15**, 225–239.

Asheshov, I. N. (1924). *J. infect. Dis.*, **34**, 534–542.

Asheshov, I. N., Asheshov, I., Khan, S., Lahiri, M. N., and Chatterjee, S. K. (1933). *Ind. J. med. Res.*, **20**, 1127–1157.

Azurin, J. C., Kobari, K., Barua, D., Alvero, M., Gomez, C. Z., Dizon, J. J., Nakano, E., Suplido, R., and Ledesma, L. (1967). *Bull. Wld Hlth Org.*, **37**, 745–749.

Bart, K. J., Huq, Z., Khan, M., and Mosley, W. H. (1970). *J. infect. Dis.*, **121**, (Suppl.), 517–524.

Barua, D. (1963). *Nature, Lond.*, **200**, 710–711.

Barua, D. (1974). *In* "Cholera" (D. Barua and W. Burrows, Eds), pp. 85–126 W. B. Saunders, Philadelphia.

Barua, D., and Chatterjee, S. N. (1964). *Ind. J. med. Res.*, **52**, 828–830.

Barua, D., and Gomez, C. Z. (1967). *Bull. Wld Hlth Org.*, **37**, 800–803.

Barua, D., and Mukherjee, A. C. (1963). *Bull. Cal. Sch. trop. Med.*, **11**, 85–86

Barua, D., and Mukherjee, A. C. (1964). *Bull. Cal. Sch. trop. Med.*, **12**, 147–149.

Barua, D., and Sack, R..B. (1964). *Ind. J. med. Res.*, **52**, 855–866.

Barua, D., and Watanabe, Y. (1972). *J. Hyg.*, **70**, 161–169.

Barua, D., Mukherjee, A. C., and Sack, B. (1964). *Bull. Cal. Sch. trop. Med.*, **12**, 55–56.

Basu, S., and Mukerjee, S. (1968). *Experentia*, **24**, 299–300.

Basu, S., and Mukerjee, S. (1970). *Bull. Wld Hlth Org.*, **43**, 509–512.

Basu, S., Bhattacharyya, P., and Mukerjee, S. (1966). *Bull. Wld Hlth Org.*, **34**, 371–378.

Benenson, A. S., Islam, M. R., and Greenough III, W. B. (1964). *Bull. Wld Hlth Org.*, **30**, 827–831.

Benenson, A. S., Saad, A., and Paul, M. (1968). *Bull. Wld Hlth Org.*, **38**, 267–276.

Bhaskaran, K., and Gorill, R. H. (1957). *J. gen. Microbiol.*, **16**, 721–729.

Bhaskaran, K., and Sinha, V. B. (1971). *J. gen. Microbiol.*, **69**, 89–97.

Bhaskaran, K., Iyer, C. S., Khan, A. W., and Vora, V. C. (1974). *Antibiotics and Chemotherapy*, **6**, 375–378.

Bhattacharyya, F. K. (1970). D.Phil. Thesis, University of Calcutta.

Bhattacharyya, F. K. (1975). *Med. Microbiol. Immunol.*, **162**, 29–41.

Bhattacharyya, F. K., and Mukerjee, S. (1974). *Ann. Microbiol.*, **125A**, 167–181.

Bhattacharyya, F. K., Narayanswami, A., and Mukerjee, S. (1970). Proc. Ind. Counc. Med. Res. Seminar on Immunity and Immunoprophylaxis in Cholera, Calcutta. *Ind. Counc. med. Res. Tech. Rpt. Ser. No. 9*, 55–57.

Bordet, J., and Ciuca, M. (1921). *C. R. Soc. Biol.*, **84**, 276–279.

Breed, R. S., Murray, E. G. D., and Smith, N. R. (1957). "Bergey's Manual of Determinative Bacteriology". 7th edn. Williams & Wilkins, Baltimore.

Brenner, S., and Horne, R. W. (1959). *Biochem. Biophys. Acta*, **34**, 103–110.

Burnet, F. M. (1927). *Brit. J. exp. Path.*, **8**, 121.

Burnet, F. M. (1930). *J. Path. Bact.*, **33**, 647–664.

Burnet, F. M. (1933). *J. Path. Bact.*, **36**, 317–318.

Burnet, F. M. (1934). *J. Path. Bact.*, **38**, 285–299.

Burnet, F. M., and McKie, M. (1930). *J. Path. Bact.*, **33**, 637–646.

Chakrabarty, A. N., and Dastidar, S. G. (1974). *J. gen. Microbiol.*, **80**, 339–361.

Chakrabarty, A. N., Adhya, S., Basu, J., and Dastidar, S. G. (1970). *Inf. Imm.*, **1**, 293–299.

Chakrabarty, A. N., Dastidar, S. G., and Adhya, S. (1971a). Proc. Ind. Counc. Med. Res. Seminar on Immunity and Immunoprophylaxis in Cholera, Calcutta. *Ind. Counc. med. Res. Tech. Rpt. Ser. No.* 9, 190–193.

Chakrabarty, A. N., Sil, J., and Mukerjee, S. (1971b). Proc. Ind. Counc. Med. Res. Seminar on Immunity and Immunoprophylaxis in Cholera, Calcutta. *Ind. Counc. med. Res. Tech. Rpt. Ser. No.* 9, 161–167.

Chatterjee, S. N., Das, J., and Barua, D. (1965). *Ind. J. med. Res.*, **53**, 934–937.

Chatterjee, S. N., and Maiti, M. (1971). *J. gen. Virol.*, **13**, 327–330.

Chun, D., Chung, J. K., Lew, S. T., and Chaug, M. W. (1970). *J. Korean Med. Assn.*, **13**, 53–58.

Craigie, J., and Yen, C. H. (1938). *Can. Publ. Hlth J.*, **29**, 448–463.

Das, J., and Chatterjee, S. N. (1966). Proc. Sixth Int. Congr. for Electron Microscopy, Kyoto, Japan, pp. 139–140.

De Lahiri, S. N., Chaudhury, P. K., Ghosh, M. L., and Mondal, A. (1958). *Ind. J. med. Res.*, **46**, 351–358.

Dutta, N. K., Sil, J., Sanyal, S. C., and Mukerjee, S. (1972). Proc. Ind. Counc. med. Res. Seminar on Immunity and Immunoprophylaxis in Cholera, Calcutta. *Ind. Counc. Med. Res. Tech. Rpt. Ser. No.* 9, 226–231.

Farkas-Himsley, H., and Seyfried, P. L. (1962). *Nature, Lond.*, **193**, 1193–1194.

Farkas-Himsley, H., and Seyfried, P. L. (1963). *Can. J. Microbiol.*, **9**, 339–343.

Feeley, J. C. (1969). *J. Bact.*, **99**, 645–649.

Feeley, J. C., and Pittman, M. (1963). *Bull. Wld Hlth Org.*, **28**, 347–356.

Felsenfeld, O. (1963a). *Bull. Wld Hlth Org.*, **28**, 289–296.

Felsenfeld, O. (1963b). Pers. comm.

Felsenfeld, O. (1964). *Bacteriol. Rev.*, **28**, 72–86.

Felsenfeld, O., Soman, D. W., Young, V. M., Yoshida, T., Waters, T., and Ishihea, S. J. (1951). *Proc. Soc. exp. Biol.*, **77**, 284–286.

Finkelstein, R. A. (1962). *J. Immunol.*, **89**, 264–271.

Finkelstein, R. A., and La Brec, E. H. (1959). *J. Bact.*, **78**, 886–889.

Finkelstein, R. A., and Mukerjee, S. (1963). *Proc. Soc. exp. Biol.*, **112**, 355–359.

Gallut, J. (1962). *Ann. Inst. Pasteur*, **102**, 309–327.

Gallut, J. (1965). Proc. Cholera Res. Symp., Honolulu. Publ. Hlth Serv. Publication No. 1328, Washington, 235–243.

Gallut, J., and Nicolle, P. (1963). *Bull. Wld Hlth Org.*, **28**, 389–393.

Gan, K. H., and Tjia, S. K. (1963). *Am. J. Hyg.*, **77**, 184–186.

Gangarosa, E. J., De Witt, W. E., Feeley, J. C., and Adams, M. R. (1970). *J. infect. Dis.*, **121** (Suppl.), 536–543.

Gardner, A. B., and Venkatraman, R. V. (1935). *J. Hyg.*, **35**, 262–282.

Ghosh, S. N., and Mukerjee, S. (1960). *Ann. Biochem. exp. Med.*, **20**, 251–256.

Ghosh, S. N., and Mukerjee, S. (1961). *Ann. Biochem. exp. Med.*, **21**, 151–156.

Gielen, J. M. Th., De Jongh, R. T., and Weiland, H. (1972). Bacterial Disease Branch Epidemiology Programme, CDC. *Morbid. Mortal.*, **21**, 238–239.

Givental, N. I., Gritzenko, A. N., Podosinuskova, L. S., and Ermolieva, Z. V. (1974). *Zh. Mikrobiol. Epidemiol. Immunol.*, No. 2, 54–57.

Goebel, W. F. (1950). *J. exp. Med.*, **92**, 527–543.

Goebel, W. F., and Jesaitis, M. A. (1953). *Ann. Inst. Pasteur*, **84**, 66–72.

Goebel, W. F., and Perlman, G. E. (1949). *J. exp. Med.*, **89**, 479–489.

Gotschlich, F. (1906). *Z. Hyg. Infekt-Kr.*, **53**, 281–304.

Gratia, A. (1922). *J. exp. Med.*, **35**, 281–304.
Gratia, A. (1940). *C. R. Soc. Biol.*, **133**, 702–703.
Greig, E. D. W. (1914). *Ind. J. med. Res.*, **2**, 623–647.
Gruber, M., and Durham, H. E. (1896). *Münch. Med. Wschr.*, **43**, 285.
Guha Roy, U. K., and Mukerjee, S. (1959). *Ann. Biochem.*, **19**, 125–130.
Guhathakurta, B., and Dutta, G. S. (1974). *Appl. Microbiol.*, **27**, 634–639.
Heiberg, B. (1934). *C. R. Soc. Biol. (Paris)*, **115**, 984–986.
D'Herelle, F. (1926). "The Bacteriophage and its Behaviour". Williams and Wilkins, Baltimore.
Hershey, A. D. (1955). *Virology*, **1**, 108–127.
Hershey, A. D., and Chase, M. (1952). *J. gen. Physiol.*, **36**, 39–56.
Hugh, R., and Feeley, J. C. (1972). *Int. J. syst. Bact.*, **22**, 123.
Jackson, G. D. F., and Redmond, J. W. (1971). *FEBS Letters*, **13**, 117–120.
Jann, B., Jann, K., and Beyaert, G. O. (1973). *Eur. J. Biochem.*, **37**, 531–534.
Jayawardene, A., and Farkas-Himsley, H. (1968). *Nature, Lond.*, **219**, 79–80.
Jesaitis, M. A., and Goebel, W. F. (1952). *J. exp. Med.*, **96**, 409–423.
Kauffman, F. (1950). *Acta path. microbiol. scand.*, **27**, 283–299.
Koch, R. (1884). *Dtsch. med. Wschr.*, **10**, 725–728.
Kraus, R., and Prantschoff, A. (1906). *Wien. k. Wschr.*, **19**, 299–300.
Lanni, F., and Lanni, Y. T. (1956). *Bact. Proc.* 51.
Lark, K. G., and Adams, M. H. (1953). *Cold Spring Harbour Symposia Quant. Biol.*, **18**, 171–183.
Levine, P., and Frisch, A. N. (1934). *J. exp. Med.*, **59**, 213–228.
Liu, P. U. (1959). *J. Infect. Dis.*, **104**, 238–252.
Loghem, J. J. Van (1926). *Zbl. Bakt. Abt. I Orig.*, **100**, 19–21.
Luria, S. E. (1945). *Genetics*, **30**, 84–99.
Luria, S. E., and Delbruck, M. (1943). *Genetics*, **28**, 491–511.
Lyles, S. T., and Gardner, E. W. (1958). *J. infect. Dis.*, **103**, 67–74.
Maiti, M., and Chatterjee, S. N. (1969a). *Bull. Cal. Sch. trop. Med.*, **77**, 119–120.
Maiti, M., and Chatterjee, S. N. (1969b). *Bull. Cal. Sch. trop. Med.*, **77**, 77–79.
Malizia, W. F. (1954). *U.S. Armed Forces Med. J.*, **5**, 1528–1530.
Miller, E. H., and Goebel, W. F. (1949). *J. exp. Med.*, **90**, 255–265.
Miller, E. H., and Goebel, W. F. (1952). *J. exp. Med.*, **96**, 425.
Misra, S. B., and Shrivastava, D. L. (1961). *Ind. J. med. Res.*, **49**, 183–188.
Misra, S. B., Pant, K. D., and Shrivastava, D. L. (1961). *Ind. J. med. Res.*, **49**, 963–973.
Monsur, K. A., Rizvi, S. S. H., Huq, M. I., and Benenson, A. S. (1965). *Bull. Wld Hlth Org.*, **32**, 211–216.
Monsur, K. A., Rahmann, M. A., Hcia, F., Islam, M. N., Northrup, R. N., and Hirschhorn, N. (1970). *Bull. Wld Hlth Org.*, **42**, 723–732.
de Moor, C. E. (1949). *Bull. Wld Hlth Org.*, **2**, 5–17.
de Moor, C. E. (1963). *Trop. Georg. Med.*, **18**, 97–107.
Mukerjee, S. (1959). *Ann. Biochem.*, **19**, 9–12.
Mukerjee, S. (1960). *Ann. Biochem.*, **20**, 182–203.
Mukerjee, S. (1961a). *Ann. Biochem.*, **21**, 257–264.
Mukerjee, S. (1961b). *Ann. Biochem.*, **21**, 265–266.
Mukerjee, S. (1961c). *Ann. Biochem.*, **21**, 267–270.
Mukerjee, S. (1961d). *Ann. Biochem.*, **22**, 1–4.
Mukerjee, S. (1962). *Ann. Biochem.*, **22**, 5–8.
Mukerjee, S. (1963a). *Bull. Wld Hlth Org.*, **28**, 333–336.

Mukerjee, S. (1963b). *Bull. Wld Hlth Org.*, **28**, 337–345.

Mukerjee, S. (1964a). *Ind. J. med. Res.*, **52**, 331–354.

Mukerjee, S. (1964b). *Ind. J. med. Res.*, **52**, 771–776.

Mukerjee, S. (1965). Proc. Cholera Res. Symp., Honolulu. U.S. Publ. Hlth Ser. Publication No. 1328, 9–16.

Mukerjee, S. (1967). *Ind. J. med. Res.*, **55**, 308–313.

Mukerjee, S. (1973). *In* "Lysotypie, Infektionskrankheiten und ihre Erreger" (H. Rische, Ed.). VEB Gustav Fischer Verlag, Jena.

Mukerjee, S. (1974). *In* "Cholera". (D. Barua and W. Burrows, Eds). W. B. Saunders Company, Philadelphia.

Mukerjee, S., and Guha Roy, U. K. (1961a). *Ann. Biochem.*, **21**, 129–132.

Mukerjee, S., and Guha Roy, U. K. (1961b). *J. Bact.*, **81**, 830–831.

Mukerjee, S., and Guha Roy, U. K. (1962). *Br. med. J.*, **1**, 685–687.

Mukerjee, S., and Takeya, K. (1974). *In* "Cholera". pp. 61–84. (D. Barua and W. Burrows, Eds). W. B. Saunders Company, Philadelphia.

Mukerjee, S., Basu, S., and Bhattacharyya, P. (1965). *Br. med. J.*, **II**, (3), 837–839.

Mukerjee, S., Guha, D. K., and Guha Roy, U. K. (1957). *Ann. Biochem.*, **18**, 163–176.

Mukerjee, S., Guha Roy, U. K., and Rudra, B. C. (1960). *Ann. Biochem.*, **20**, 182–203.

Mukerjee, S., Guha Roy, U. K., Guha, D. K., and Rudra, B. C. (1959). *Ann. Biochem.*, **19**, 115–124.

Neogy, K. N., and Sanyal, S. N. (1966). *Bull. Cal. Sch. trop. Med.*, **14**, 94–95.

Neoh, S. H., and Rowley, D. (1970). *J. infect. Dis.*, **121**, 505–513.

Newman, E. S., and Eisenstark, A. (1964). *J. infect. Dis.*, **114**, 217–225.

Nicolle, P., Gallut, J., and Le Minor, A. (1960). *Ann. Inst. Pasteur*, **99**, 664–671.

Nicolle, P., Gallut, J., Ducrest, P., and Quiniou, J. (1962). *Rev. Hyg. Med. Soc.*, **10**, 91–126.

Nobechi, N. (1964). *J. Jap. Med. Ass.*, **52**, 247–251.

Ohen, W. K. (1932). *Proc. Soc. exp. Biol.*, **29**, 1160–1162.

Ohen, W. K. (1933). *Proc. Soc. exp. Biol.*, **30**, 887–891.

Otto, R., and Winkler, W. F. (1922). *Dtsch. med. Wschr.*, **48**, 382–384.

Pant, K. D. (1968). *Ind. J. Microbiol.*, **8**, 1–10.

Pesigan, T. P., Cromez, C. L., Gaeros, D., and Barna, D. (1967). *Bull. Wld Hlth Org.*, **37**, 795–797.

Pasricha, C. L., deMonte, A. J. H., and Gupta, S. K. (1936). *Ind. med. Gaz.*, **71**, 194–196.

Pasricha, C. L., Lahiri, M. N., deMonte, A. J. H., and Gupta, S. K. (1936). *Ind. med. Gaz.*, **76**, 218–219.

Pollitzer, R. (1959). "Cholera". WHO Monograph Series No. 43, WHO, Geneva.

Redmond, J. W. (1975). *FEBS Letters*, **50**, 147–149.

Redmond, J. W., Korsch, M. J., and Jackson, G. D. F. (1973). *Aust. J. exp. biol. Sci.*, **51**, 229–235.

Rizvi, S., and Benenson, A. S. (1966). *Bull. Wld Hlth Org.*, **35**, 675–680.

Roy, C., and Mukerjee, S. (1966). *J. gen. App. Microbiol.*, **12**, 179–190.

Roy, C., Mridha, K., and Mukerjee, S. (1965). *Proc. Soc. expt. Biol. Med.*, **119**, 893–896.

Sack, R. B., and Miller, C. E. (1969). *J. Bact.*, **99**, 688–695.

Sakazaki, R. (1970). *WHO Public Health Paper*, **40**, 33–37.

Sakazaki, R. (1968). *Jap. J. Med. Sci. Biol.*, **21**, 359–362.

Sakazaki, R., and Shimada, T. (1975). Proc. Int. Symp. on Vibrionaceae and other Agents of Enterotoxicosis, Kosice, 16.

Sakazaki, R., and Tamura, K. (1971a). Unpublished assignment report to WHO, Geneva.

Sakazaki, R., and Tamura, K. (1971b). *Jap. J. med. Sci. Biol.*, **24**, 93–100.

Sakazaki, R., Iwanani, S., and Fukami, H. (1963). *Jap. J. Med. Sci. Biol.*, **16**, 161–188.

Sakazaki, R., Iwanami, R., and Tamura, K. (1968). *Jap. J. med. Sci. Biol.*, **21**, 313–324.

Sakazaki, R., Tamura, K., Gomez, C. Z., and Sen, R. (1970). *Jap. J. med. Sci. Biol.*, **23**, 13–20.

Sanyal, S. C., and Sakazaki, R. (1974). *Ind. J. Microbiol.*, **14**, 150–158.

Sanyal, S. C., Sil, J., Dutta, N. K., and Mukerjee, S. (1972). *Ind. J. med. Res.*, **60**, 1138–1139.

Sertic, V., and Boulgakov, N. A. (1935). *C. R. Soc. Biol.*, **119**, 1270–1272.

Shimada, T., and Sakazaki, R. (1973). *Jap. J. med. Sci. Biol.*, **26**, 155–160.

Shrivastava, D. L. (1965). Proc. Cholera Res. Symp., Honolulu. Pub. Hlth. Serv. Publication No. 1328, Washington, 244–247.

Shrivastava, D. L., and White, P. B. (1947). *Ind. J. Med. Res.*, **35**, 117–130.

Sil, J., Dutta, N. K., Sanyal, S. C., and Mukerjee, S. (1972). *Ind. J. Microbiol.*, **12**, 110–112.

Sil, J., Dutta, N. K., Sanyal, S. C., and Mukerjee, S. (1974). *Ind. J. med. Res.*, **62**, 15–21.

Smith, H. L. (1974). *Appl. Microbiol.*, **27**, 375–378.

Stocker, B. A. D. (1955). *J. gen. Microbiol.*, **12**, 375–381.

Takeya, K. (1967). *Jap. J. Trop. Med.*, **8**, 5–9.

Takeya, K. (1974). *In* "Cholera". (D. Barua and W. Burrows, Eds), pp. 74–76. W. B. Saunders Company, Philadelphia.

Takeya, K., and Shimodori, S. (1969). *J. Bact.*, **99**, 339–340.

Takeya, K., and Shimodori, S. (1963). *J. Bact.*, **85**, 957–958.

Takeya, K., Shimodori, S., and Gomez, C. Z. (1967). *Bull. Wld Hlth Org.*, **37**, 806–810.

Takeya, K., Shimodori, S., and Zinnaka, Y. (1965a). *J. Bact.*, **90**, 824–825.

Takeya, K., Zinnaka, Y., Shimodori, S., Nakayama, Y., Amko, K., and Lida, K. (1965b). *In* "Proc. Cholera Res. Symp.", Honolulu, pp. 24–29.

Tanamal, S. J. W. (1938). *Ned. T. Geneesek.*, **82**, 1370–1375.

Tanamal, S. J. W. (1959). *Am. J. Trop. Med.*, **8**, 72–73.

Vasil, M. L., Holmes, R. K., and Finkelstein, R. A. (1974). *Inf. Imm.*, **9**, 195–197.

Vieu, J. F., Nicolle, P., Gallut, J., and Ducrest, P. (1965). Proc. Cholera Res. Symp., Honolulu, Hawaii. U.S. Govt. Printing Office, Washington, pp. 34–36.

Wahba, A. H., and Takla, U. (1962). *Bull. Wld Hlth Org.*, **26**, 306–308.

White, P. B. (1935). *J. Hyg.*, **35**, 347–353.

White, P. B. (1940). *J. Path. Bact.*, **50**, 165–166.

Winkle, S., Refai, M., and Rohde, R. (1972). *Ann. Inst. Pasteur*, **123**, 775–781.

WHO Report Series No. 52 (1952). Expert Committee on Cholera, First Report, Geneva, p. 7.

Zinder, N. D., and Lederberg, J. (1952). *J. Bact.*, **64**, 679–699.

Zinnaka, Y., Oshikala, H., and Takeya, K. (1966). *Med. Biol.*, **72**, 220–223.

Zinnaka, Y., Shimodori, S., and Takeya, K. (1964). *Jap. J. med. Microbiol.*, **8**, 97–103.

CHAPTER VI

# Vibriocin Typing

H. BRANDIS

*Institute of Medical Microbiology and Immunology, University of Bonn*

## I. INTRODUCTION

*Vibrio cholerae* and *V. cholerae* biotype *eltor* can be differentiated by serological and phage typing. However, these two methods have their limitations, since with the serological method only two main types (Ogawa and Inaba) can be detected. With phage typing, five types of *V. cholerae* and six types of *V. cholerae* biotype *eltor* have been identified (cf. this Volume, Chapter IV). Vibriocin typing could provide an additional tool for epidemiological purposes. With an appropriate system, it should be possible to subdivide a serological type or a phage type by means of vibriocin typing.

It must be emphasised that no standardised bacteriocin typing system which could be used in laboratory routine work presently exists for the differentiation of *V. cholerae* and *V. cholerae* biotype *eltor*. In this Chapter, the difficulties in dealing with vibriocin and the preliminary trials of several authors in using vibriocin typing are summarised in order to present the current state of research on which further investigations may be built. Bacteriocins of *V. cholerae* are difficult to detect. Nicolle *et al.* (1962) and Barua (1963) failed to demonstrate such a bacteriocin. However, different kinds of bacteriocins of *V. cholerae* may exist. Baskaran (1960, 1964), working with genetic recombination of strains with and without fertility factors, P+ and P−, found that P+ strains produced a bacteriocin-like substance which was distinct from bacteriophage. This inhibited the

growth of P⁻ strains. With Fredericq's chloroform method, the antibacterial substance could not be demonstrated, but this was possible if a
drop of the producing P⁺ culture was placed on a soft agar layer containing
sensitive indicator cells in appropriate density. These observations suggest
that the bacteriocin activity is cell-bound. In fact, Takeya and Shimodori
(1969) detected such an antibiotic substance by a "lacuna" forming
method. When approximately 50–100 bacteriocinogenic bacteria were
mixed with a proper concentration of indicator cells (VC 154 strain of
Asiatic cholera) in soft (0·3%) Brain Heart Infusion Agar overlay, "lacu-
nae" (i.e. plaque-like clearances around the microcolonies of the producing
cells) could be observed after incubation at 37°C for 6 h followed by
holding at room temperature. Bacteriocins were detected in two P⁺ *V.
cholerae* strains and five of 410 strains of *Vibrio cholerae* biotype *eltor*. It
was not possible to separate the bacteriocin from living bacteriocinogenic
cells. Induction with Mitomycin C was likewise unsuccessful. Growth
inhibition zones using the stab culture method could only be found when
the producer cells were not killed by chloroform vapour.

With the electron microscope no phage-like structures were detected.
Release of bacteriocin from the cells by cold shock has not been investigated.

Farkas-Himsley and Seyfried (1962) found another antibiotic principle
designated vibriocin in a strain of *V. cholerae*. This has mainly been
investigated in one single system.

## II. PRODUCTION OF VIBRIOCIN

As a vibriocin producing strain, Farkas-Himsley and Seyfried (1962)
used the Streptomycin sensitive *V. cholerae* ATCC 9168. This produced
vibriocin only after induction by anaerobiosis (e.g. thioglycollate (0·15%
final concentration in nutrient broth)), ultraviolet irradiation, or cold
shock. Later, Jayawardene and Farkas-Himsley (1969a) showed that also
Mitomycin C (0·3 μg/ml for 10 min) induced vibriocin production. A
combination of Mitomycin C and cold shock was optimal for getting high
titres. The conditions for vibriocin induction were therefore closely
related to induction of bacteriophages. No infective phage particles,
however, could be detected by the authors. After induction with Mitomycin C, the growth of *V. cholerae* still increased for approximately 1 hour,
then stopped or decreased. This was due to lysis of the cells. Therefore,
the vibriocin production was characterised by a lethal biosynthesis. The
efficiency of induction was related to the growth phase of the vibriocinogenic bacteria. A complex medium seemed essential for production,
which was not observed in synthetic media (Jayawardene and Farkas-

Himsley, 1969a). Vibriocin can be detected 45 min after induction and then reaches values of $8 \times 10^6$ lethal units/ml (Jayawardene and Farkas-Himsley, 1970a).

For the detection of vibriocin activity a Streptomycin resistant mutant of strain 9168 was used. This showed a high sensitivity under aerobic, but not under anaerobic conditions, (Farkas-Himsley and Seyfried, 1962, 1963a, b; Jayawardene and Farkas-Himsley, 1969b). The broth culture of the producer strain was centrifuged and the supernatant filtered through Gradocol membranes or Millipore filters (0·45 $\mu$m pore size). Three methods were used to demonstrate bactericidal effect:

(a) by placing a drop of the supernatant containing the vibriocin on the surface of a nutrient agar plate seeded with the indicator strain,

(b) by a turbidimetric measuring apparatus. The broth culture of the indicator strain was diluted to a given density, the vibriocin-containing supernatant added and turbidimetric measurements were carried out. The vibriocin titre (arbitrary units per ml) was expressed as the reciprocal value of the highest dilution of the supernatant giving growth inhibition. Individual batches of vibriocin showed a marked variation in titre,

(c) by demonstrating inhibition zones around single colonies of the producer strains using a modified plate overlay method of Fredericq (1954).

The number of lethal units was determined by incubation of a saline suspension of the indicator cells in the exponential phase with a sample of vibriocin at 37°C. After 30 min, the remaining fraction of the original indicator-cell population was determined by viable count and the concentration of lethal units per millilitre calculated. The colourimetric modification of this assay as proposed by Shannon and Hedges (1970) proved simple, rapid and reliable.

A rapid quantitative assay for vibriocin activity, based on the circumstance that sensitive cells treated with vibriocin absorb less ultraviolet light (260 nm) has been worked out by Krol and Farkas-Himsley (1970) (for techniques see Mayr-Harting et al., 1972).

## III. HOST RANGE

Farkas-Himsley and Seyfried (1963b) found an inverse relationship between sensitivity to vibriocin and sensitivity to Streptomycin. Among 23 Streptomycin resistant strains, 19 (83%) were sensitive to vibriocin, whereas of 17 Streptomycin sensitive strains, 13 (78%) were resistant. Lang et al. (1968) found that vibriocin from strain RC 10425 was active on 10 of 28 V. cholerae strains when tested with a modification of Wahba's method. However, vibriocin is not species specific. Farkas-Himsley and Seyfried (1963b) among eight strains of Escherichia coli found four sensitive to vibriocin (three strains were Streptomycin resistant). Strains of *Pseudomonas aeruginosa* and *P. fluorescens* were also sensitive to vibriocin (and resistant to Streptomycin). Datta and Prescott (1969a) confirmed that vibriocin inhibits a wide range of enterobacteriae. Vibriocins from two strains inhibited four of six *E. coli* strains, all four *Enterobacter aerogenes*, one *E. cloacae* tested, one of two *Klebsiella pneumoniae*, both *Proteus vulgaris* examined, three of four *Shigella flexneri*, two of three *S. boydii* and the one *S. dysenteriae* strain tested.

Moreover, vibriocin preparations $(3 \times 10^8 \text{ LU/ml})$ have a marked cytotoxic effect on actively multiplying HeLa and L-929 mouse fibroblast cells in tissue cultures (Farkas-Himsley, 1974), although the possibility of some other toxic agent being present in the supernatants has not been ruled out.

## IV. PHYSICAL AND MORPHOLOGICAL PROPERTIES

Vibriocin from strain ATCC 9168 is inactivated by trypsin (0·025% w/v) at 37°C (pH 8) and by other proteolytic enzymes. At pH levels ranging from 1·3 to 12·5 no inactivation occurred. Exposure to 56°C for 3 or boiling for 1 min did not destroy the activity, but it disappeared after 30 min boiling (Farkas-Himsley and Seyfried, 1963a). Farkas-Himsley and Seyfried (1965) fractionated the supernatant of a vibriocin producing culture by sucrose gradient centrifugation. The active fractions contained protein, nucleic acids and phospholipids. This was interpreted to indicate that vibriocins structurally may resemble defective phages. In subsequent studies, Jayawardene and Farkas-Himsley (1968), using electron microscopy, showed that vibriocin after partial purification by fractionation on Sephadex G 200 consisted of double hollow cylinders. These were 1100 Å long and had an outer sheath and an inner core. The width of the sheath was approximately 240 Å and the diameter of the core 90–100 Å. Contracted and extended sheaths with empty or full inner cores were observed. Lang *et al.* (1968) confirmed these observations and suggested that the particles could be identical with tail portions of *V. cholerae* bacteriophage. Parker *et al.* (1970) tested 27 strains of *V. cholerae* induced with Mitomycin C. Twelve strains produced phage tail-like material and one strain typical bacteriophages. Both P+ and P− strains generated phage tails. Thus, no correlation was seen between P+ character and the presence or absence of phage tails. However, the authors did not investigate the vibriocin activity of their preparations. In cells producing vibriocin, the morphological changes resemble those in defective lysogenic cells producing phage components after induction (Farkas-Himsley *et al.*, 1971).

## V. MODE OF ACTION

The lethal action of vibriocin on sensitive cells follows single hit kinetics (Jayawardene and Farkas-Himsley, 1969b). Adsorption on sensitive cells is temperature independent but the lethal action occurs only in an appropriate narrow temperature range and requires oxidative phosphorylation. The cells are most sensitive in their exponential growth phase (Jayawardene and Farkas-Himsley, 1969b).

As is the case with some colicins, the lethal action of vibriocin is reversed by trypsin added within 7–10 min after adsorption (Jayawardene and Farkas-Himsley, 1970b). Jayawardene and Farkas-Himsley (1970b) found that vibriocin does not act by lysing sensitive bacteria, but by damaging the cytoplasmic membrane. This was indicated by leakage of u.v.-absorbing (260 nm) material and efflux of radioactive $^{42}K$ followed by the inhibition of the synthesis of macromolecules (Krol and Farkas-Himsley, 1972). Vibriocin stops almost all DNA synthesis and DNA degradation occurs after 10 min. RNA synthesis continues for a limited time at a reduced rate. Protein synthesis is least inhibited. Chloramphenicol (5 $\mu$g/ml) protects sensitive cells against the bactericidal action of vibriocin. Therefore, protein synthesis seems necessary for vibriocin action. Sensitive cells in the stationary growth phase or under anaerobic conditions are tolerant to vibriocin (Krol and Farkas-Himsley, 1971).

Vibriocinogenic bacteria adsorb vibriocin, but are immune to its action. Upon induction, however, immunity to vibriocin is lost (Jayawardene and Farkas-Himsley, 1970a).

## VI. TYPING PROCEDURES

The intention of Wahba (1965) was to determine whether a relationship indicated by vibriocin production exists between *V. cholerae* and *V. cholerae* biotype *eltor*. For these studies, Nutrient Agar No. 2 (Oxoid) was chosen. The method of revealing bacteriocin production was as follows: the strain to be tested was stabbed on to plates of nutrient agar, incubated over night at 37 °C, and then left at 4 °C for 6 h, when the bacterial growth was killed by chloroform vapour (1 h). Thereafter, the plates were flooded with a 4 hours' old broth culture of the indicator strain. The plates were maintained first at 4 °C for 18 h and then incubated for 8 h at 37 °C. Finally they were examined for the presence of inhibition zones around the macrocolonies of the test strain. Eight *V. cholerae* and eight *V. cholerae* biotype *eltor* strains were used for a chequerboard experiment in which all 16 strains were tested for production and sensitivity to vibriocin. One *V. cholerae* strain and two *V. cholerae* biotype *eltor* strains produced vibriocin active on some of the strains. Two *V. cholerae* and two *V. cholerae* biotype *eltor* strains were sensitive to the vibriocins of the producer strains.

Datta and Prescott (1969a) found several methods used for detection of bacteriocins of vibrios to be unsatisfactory. In their hands, Wahba's method gave variable results. The authors modified this method in order to get more clear-cut results. Instead of the stab method, they studied diffusion from circular reservoirs in nutrient agar filled with melted agar containing the cells of the test strain ($10^9$/ml). After incubation at 37 °C for 48 h and 6 h at 4 °C, the bacterial growth was killed by chloroform vapour (1 h). Then, an overlay of nutrient agar with indicator cells was poured on the plates. The plates were held for 18 h at 4 °C and incubated for 24 h at 37 °C before reading. With this modification, the inhibition zones were small (1–3 mm). Bacteriophage action in the case of inhibition was excluded.

With the same method Datta and Prescott (1969b) examined the possibility of using vibriocins to differentiate *V. cholerae* from *V. cholerae* biotype *eltor*. Two strains of *V. cholerae* (NIHR41 and NIHR35A3) were quite active, and a third strain weak vibriocin producers against strains of both *V. cholerae* and *V. cholerae* biotype *eltor*. *V. cholerae* NIHR41 was active against eight *V. cholerae* and 16 *V. cholerae* biotype *eltor* strains out of 42 strains tested as indicators. *V. cholerae* NIHR35A3 was active against 12 *V. cholerae* and 11 *V. cholerae* biotype *eltor* strains. Identical inhibition patterns were found against six *V. cholerae* and nine *V. cholerae* biotype *eltor* strains. Sixteen *V. cholerae* biotype *eltor* strains were tested against *V. cholerae* and *V. cholerae* biotype *eltor* indicator strains. Thirteen of these 16 strains produced an inhibition on one or more of 13 out of 14 *V. cholerae* and *V. cholerae* biotype *eltor* indicators. Six from eight of the more active strains showed an activity on eight of 15 *V. cholerae* indicator strains. It was, therefore, not possible to use vibriocins as a taxonomic tool in differentiating *V. cholerae* from *V. cholerae* biotype *eltor*. The question whether vibriocin production could be used as an epidemiological marker was not examined by the authors. Preliminary purification indicated that the vibriocin of *V. cholerae* NIHR41 belonged to the high molecular weight bacteriocins.

Chakrabarty *et al.* (1970), in contrast to the above investigators, demonstrated abundant vibriocin production by conventional techniques and various media. For typing, Tryptic Soy Agar Base Medium (DIFCO) was adopted. The authors tested the influence of various factors on vibriocin production and found that some were essential: Citrate-phosphate buffer at 0·5–0·7% concentration, pH 7·5–7·6 and a cold shock for 18 h.

Chakrabarty and Dastidar (1976) recommended the following medium:

*Base:* Tryptic Soy Broth (DIFCO) 15 g, Agar (DIFCO) 15 g, distilled water to make up the volume to 955 ml, mixed, dissolved by steaming for 10 min and autoclaved at 115 °C for 12 min. *Solutions:* (A) Trisodium citrate (Analar) 5 g in 20 ml distilled water; (B) $K_2HPO_4$ (anhydrous, Analar) 5 g in 20 ml distilled water. (C) $NH_4Cl$ (anhydrous, Analar) 30 mg in 5 ml distilled water. Solutions (A), (B), and (C) are steamed separately for 30 min, then mixed together and added to the base at 50 °C, thoroughly mixed and plates are poured by adding 20 ml amounts to 9 cm diameter Petri dishes. Routine pH checks are not needed as the pH invariably lies between pH 7·5 and 7·6.

As typing procedure, the authors use the method of Abbott and Shannon (1958) and some of their indicators. The strains to be typed are grown as streaks of 5 mm width across a plate for 54 h at 37 °C followed by storage at 4 °C for 18 h. The bacterial growth is then scraped off and the residual bacteria are killed by chloroform vapour (2 h). Eight indicator strains (18 h old peptone water (0·75% Bacto-Peptone (DIFCO) and 0·5% NaCl in distilled water, pH 7·5)) cultures are then inoculated as streaks at right angles to the original streak of the test strain. Thermal treatment of the suspensions of indicator bacteria at 45 °C for 10 min may be useful for strains giving weak and inconsistent results and for some untypable strains. The plates are incubated for 18 h at 37 °C.

As indicator organisms† are used: (a) indicator strains of the *Shigella sonnei* bacteriocin typing system of Abbott and Shannon (1958) *E. coli* Row, *S. sonnei* M2/1 (mutant of *S. sonnei* 2, selected against *V. cholerae* vibriocin), *S. sonnei* M2/2 (another mutant), *S. sonnei* M56 (mutant of *S. sonnei* 56 selected against *V. cholerae* vibriocin), *S. sonnei* 56, *S. sonnei* 17, (b) other strains: *S. flexneri* 6 No. 38, *S. flexneri* Y No. 3189, *V. cholerae* No. 852, *V. cholerae* No. 541. *S. sonnei* strains No. M2/1 and No. 56, show identical reactious with *S. sonnei* strain No. M2/2 or *V. cholerae* strain No. 852 respectively. The inhibition reactions are recorded as + to + + + + corresponding to 4, 8, 12 and 16 mm width of inhibition zone. As observed with other bacteriocin typing systems, the indicator organisms sometimes show a "paradoxical reaction". This implies a central band of growth, surrounded on both sides by zones of inhibitions and then by the peripheral growth.

425 strains of *V. cholerae* were typed. Of these strains, 301 belonged to the Ogawa serotype, 123 to the Inaba serotype and one to the Hikojima serotype. Twenty-three strains were strongly haemolytic and eight had a weak haemolytic activity. Phage typing was performed with 310 strains.

† Reference strains may be obtained from Dr A. N. Chakrabarty, Indian Institute of Experimental Medicine, 4 Raja Subooh Chandra Mallik Road, Calcutta 700032, India.

With the above named indicator strains eight different vibriocin types based on different inhibition patterns were detected. Type 1 could be subdivided into three subtypes (A, B, C) and type 2 into two (A, B). Thirteen % of the strains were untypable and 7·5% gave atypical and inconstant results (unclassifiable). The distribution of the strains in the different types was as follows: Type 1A 82 (19·29%), 1B 41 (9·65%), 1C 5 (1·2%), 2A 68 (16·0%), 2B 34 (8·0%), 3 34 (8·0%), 4 17 (4·0%), 5 28 (6·58%), 6 10 (2·35%), 7 9 (2·1%), 8 9 (2·1%). No correlation between serotypes, phage types, haemolytic activity or bacteriocin types was found. This means that the typing results are independent of serotypes, phage types, and biotypes. The method can be applied to both *V. cholerae* and *V. cholerae* biotype *eltor*. In the hands of Chakrabarty *et al.* (1970), the results were consistent and reproducible. The authors pointed out that the inhibitions observed were due to bacteriocins which behaved like other known bacteriocins.

For strains which were untypable with the basic set of indicator strains, a supplementary typing set was elaborated by Chakrabarty *et al.* (1971a). With four indicator strains (*S. flexneri* Y 3189, *S. flexneri* Y 30225, *E. coli* serotype I AI, *V. cholerae* 564), it was possible to divide untypable strains into three bacteriocin types. Of 56 strains typed, 28 belonged to type 1, four to type 2 and six to type 3. Eighteen were untypable. The bacteriocin types as well as the indicator organisms are stable and are regarded as a useful tool for epidemiological purposes (Chakrabarty *et al.*, 1970, 1971a).

Kokhanovskaia *et al.* (1972) described bacteriocinogenic properties of non-agglutinating (NAG) vibrios. This feature was used by Chakrabarty *et al.* (1971b) to develop a bacteriocin-typing scheme for NAG vibrios. The media for propagating the indicator organisms and for bacteriocin typing were the same as used for bacteriocin typing of *V. cholerae* (the incubation time of donor strains was extended to 78 h in the case of weak bacteriocin production). Also the indicator strains, the typing method used, and the reading of results were the same as described by Chakrabarty *et al.* (1970). With the aid of the eight indicator strains, ten distinct bacteriocin types were found among 215 strains of NAG vibrios belonging to group I-V of Heiberg. Six vibriocin types (1, 3, 6, 7, 8, 10) were identical with those of *V. cholerae* (1A, 2B, 1B, 6, 5, 4). Type 1 comprised 42·8% type 2 24·7%, type 3 11·6%, type 4 5·1%, type 5 3·7% of the strains; 2·3% of the strains were unclassifiable.

As far as can be seen, the bacteriocin typing system of Chakrabarty *et al.* has not been used by other workers hitherto. Therefore, confirmation has to be awaited. Farkas-Himsley (1973) tested 38 *Vibrio* strains by conventional methods of cross streaking or by stabbing followed by an overlay of agar with the indicator cells. Some of the strains investigated were those already tested by Wahba (1965), Datta and Prescott (1969b) and Chakrabarty *et al.* (1970). Different patterns of inhibition were found. However, Farkas-Himsley was not able to repeat the results of the above authors. Farkas-Himsley suggested that it might be possible to test the production of vibriocin in various strains of *V. cholerae* and *V. cholerae* biotype *eltor* in fluid media by Mitomycin C induction (Jayawardene and Farkas-Himsley, 1969). On the other hand, the sensitivity of strains to vibriocins could be tested in fluid media by the leakage of ultraviolet absorbing material into the supernatant of vibriocin-sensitive cells within 20 min after addition of vibriocin (Jayawardene and Farkas-Himsley, 1970; Krol and Farkas-Himsley, 1970).

In summary, it can be stated that vibriocins are phage tail-like bacteriocins which are not easy to detect. The results of different authors indicate that there are several vibriocins with different host ranges and that *V. cholerae* or *V. cholerae* biotype *eltor* have different patterns of sensitivity to vibriocins. A standardised method for routine typing is lacking at present.

## REFERENCES

Abbot, J. D., and Shannon, R. (1958). *J. clin. Path.*, **11**, 71–77.

Barua, D. (1963). *Nature, Lond.*, **200**, 710–711.

Baskaran, K. (1960). *J. gen. Microbiol.*, **23**, 47–54.

Baskaran, K. (1964). *Bull. World Hlth Org.*, **30**, 845–853.

Chakrabarty, A. N., Adhya, S., Basu, J., and Dastidar, S. (1970). *Infect. Immun.*, **1**, 293–299.

Chakrabarty, A. N., Dastidar, S. G., and Adhya, S. (1971a). *In* "Proc. Symp. Immunity and Immunoprophylaxis in Cholera", pp. 190–193. Techn. Rep. Series no. 9, Indian Council of Medical Research, New Delhi.

Chakrabarty, A. N., Sil, J., and Mukerjee, S. (1971b). *In* "Proc. Symp. Immunity and Immunoprophylaxis in Cholera", pp. 161–167. Techn. Rep. Series No. 9, Indian Council of Medical Research, New Delhi.

Chakrabarty, A. N. and Dastidar, S. G. (1976): Personal communication to K. Takeya, Kyushu University, Fukuoka.

Datta, A., and Prescott, L. M. (1969a). *J. Bact.*, **98**, 849–850.

Datta, A., and Prescott, L. M. (1969b). *Ind. J. Med. Res.*, **57**, 1402–1408.

Farkas-Himsley, H. (1974). *IRCS Med. Sci.*, **2**, 1117.

Farkas-Himsley, H., Kormendy, A., and Jayawardene, A. (1971). *Cytobios*, **3**, 97–116.

Farkas-Himsley, H., and Seyfried, P. L. (1962). *Nature, Lond.*, **193**, 1193–1194.

Farkas-Himsley, H., and Seyfried, P. L. (1963a). *Can. J. Microbiol.*, **9**, 329–338.

Farkas-Himsley, H., and Seyfried, P. L. (1963b). *Can. J. Microbiol.*, **9**, 339–343.

Farkas-Himsley, H., and Seyfried, P. L. (1965). *Zbl. Bakt. Abt. I. Orig.*, **196**, 298–302.

Fredericq, P. (1954). *C. r. Séanc Soc. Biol.*, **148**, 399–402.

Jayawardene, A., and Farkas-Himsley, H. (1968). *Nature, Lond.*, **219**, 79–80.

Jayawardene, A., and Farkas-Himsley, H. (1969a). *Microbios*, **1**, (1B), 87–98.

Jayawardene, A., and Farkas-Himsley, H. (1969b). *Microbios*, **1**, 325–333.

Jayawardene, A., and Farkas-Himsley, H. (1970a). *Infect. Immun.*, **2**, 519–520.

Jayawardene, A., and Farkas-Himsley, H. (1970b). *J. Bact.*, **102**, 382–388.

Jayawardene, A., and Farkas-Himsley (1972). *Microbios.*, **6**, 35–46.

Kokhanovskaia, T. M., Dybtsyna, T. V., and Ermolieva, Z. V. (1972). *Antibiotiki*, **17**, 991–993.

Krol, P. M., and Farkas-Himsley, H. (1970). *Microbios*, **2**, 287–289.

Krol, P. M., and Farkas-Himsley, H. (1971). *Microbios*, **3**, 183–186.

Krol, P. M., and Farkas-Himsley, H. (1972). *Microbios*, **6**, 199–211.

Lang, D., McDonald, T. O., and Gardner, E. W. (1968). *J. Bact.*, **95**, 708–709.

Mayr-Harting, A., Hedges, A. J., and Berkeley, R. C. W. (1972). *In* "Methods in Microbiology" (J. R. Norris and D. W. Ribbons, Eds), Vol. **7A**, 315–422, Academic Press, London and New York.

Nicolle, P., Gallut, J., Ducrest, P., et Quiniou, J. (1962). *Rev. Hyg. Méd. Soc.*, **10**, 91–125.

Parker, Ch., Richardson, S. H., and Romig, W. R. (1970). *Infect. Immun.*, **1**, 417–420.

Shannon, R., and Hedges, A. J. (1970). *J. appl. Microbiol.*, **33**, 555–565.

Takeya, K., and Shimodori, S. (1969). *J. Bact.*, **99**, 339–340.

Wahba, A. H. (1965). *Bull. Wld. Hlth Org.*, **33**, 661–664.

CHAPTER VII

# Genus Staphylococcus

P. OEDING

*The Gade Institute, Department of Microbiology,*
*University of Bergen, Bergen, Norway*

## I. INTRODUCTION

After the serious outbreaks of *Staphylococcus aureus* infections in hospitals of most countries, particularly in the 1950's, the situation has been rather peaceful. Recently, however, reports indicate that hospital *S. aureus* infections are again on the increase. In addition, serious infections due to hospital-transmitted *S. epidermidis* strains have become increasingly important. The time should therefore be favourable for reviewing the methods presently at our disposal for the subdivision of staphylococci in an epidemiological situation.

No detailed collected description of the serological methods available for the classification of staphylococci has hitherto been published. Subdivision of *S. aureus* by means of its agglutinogens has been shown to work well and to have several advantages, but international co-operation resulting in an accepted standard method is still lacking. The knowledge of the wall teichoic acids and their serological characteristics has furnished another serological method which contributes to the classification of staphylococci into species and biotypes. Even the well established phage typing of *S. aureus* has been subjected to modifications. Simultaneously with an increasing incidence of human *S. aureus* infections, a deterioration in the phage typability occurred which necessitated changes in the set of phages and the procedures.

The characterisation and classification of animal *S. aureus* strains has been in the focus of interest for some years. Serotyping and phage-typing using the existing schemes supplement the biotyping of animal strains and allow further subdivision, but more specific typing methods are needed. There is also a need for typing methods which characterise *S. epidermidis* strains in such a way that they can be recognised in epidemiological investigations. The taxonomy and classification of staphylococci and micrococci are undergoing change at the present time and this must be considered in the future work on typing methods.

Three principal methods for typing staphylococci exist: Phage typing, serotyping and bacteriocin typing. Phage typing was covered by Parker (1972) in Volume 7 B of this Series and bacteriocin typing is still in its infancy. These items will therefore be treated very briefly. The serological methods are considered the main subject of this Chapter.

## II. TAXA AND IDENTIFICATION

Staphylococci are non-motile, Gram-positive, spherical bacteria usually arranged in irregular clusters. They grow readily on artificial media and a variety of metabolic activities contribute to their classification. The GC

(guanine + cytosine) content of the DNA ranges from 30 to 40 mole %
(Kocur *et al.*, 1971) and their cell walls contain characteristic structures
(see below).

In most medical diagnostic laboratories staphylococci are identified by
their growth on blood agar, followed by a tube coagulation test (Subcommittee, 1965). Strains positive in this test are designated coagulase-
positive staphylococci or *S. aureus*. Other properties such as production of
haemolysins, pigment, DNase and fermentation of mannitol are usually not
considered in the routine situation. The coagulase test is rapid and simple
but its reliability in separating *S. aureus* from *S. epidermidis* strains has
recently been questioned (Oeding, 1974). Coagulase-negative staphylococci
isolated from serious infections have been shown to possess several activities
characteristic of *S. aureus*, indicating that such strains do not belong to
*S. epidermidis* (Smith and Farkas-Himsley, 1969). In contrast, it is doubtful
whether coagulase-positive animal strains that are non-pigmented and have
a low biochemical activity should be regarded as *S. aureus* (Oeding, 1974;
Devriese and Oeding, 1975).

The separation of coagulase-negative staphylococci from micrococci is
unsatisfactory in most diagnostic laboratories. A diagnosis is made either
by rough estimation or by the ability of staphylococci to produce acid from
glucose anaerobically. The standard glucose fermentation test (Subcommittee, 1965) does, however, not recognise as staphylococci a group of
strains which grows poorly anaerobically. Modifications of the glucose
fermentation test (Evans and Kloos, 1972) and other methods (Schleifer
and Kloos, 1975b) have been described, but a satisfactory routine test for
the separation of staphylococci and micrococci is still desired.

Some laboratories subdivide coagulase-negative staphylococci into
biotypes according to the scheme of Baird-Parker (1963). It should be
remembered that today *S. epidermidis* is used both as a designation of the
recognised species subdivided into biotypes (Baird-Parker, 1974a) and as
*S. epidermidis sensu stricto*, corresponding to Baird-Parker's biotype 1
and species *S. epidermidis* of Schleifer and Kloos (1975a).

The anaerobically poorly-growing staphylococci were earlier classified
as micrococci of biotype 3 (Baird-Parker, 1963), but their DNA base
composition (Mortensen and Kocur, 1967; Kocur *et al.*, 1971), type of
peptidoglycan (Schleifer and Kandler, 1972) and cell wall teichoic acids
(Oeding and Hasselgren, 1972; Baird-Parker, 1974b; Digranes and Oeding,
1975) show that they are staphylococci. The 8th edition of Bergey's
Manual of determinative bacteriology (Baird-Parker, 1974a) classifies
them in a new species: *S. saprophyticus*. These bacteria should be properly
recognised in the diagnostic laboratory because they often cause urinary
tract infections in females (Mitchell, 1965, 1968; Roberts, 1967).

A simple, but, when used alone, not totally reliable criterion for the diagnosis of *S. saprophyticus* is its resistance to Novobiocin (Mitchell and Baird-Parker, 1967). A set of four tests, *viz.* coagulation of rabbit plasma, acid production aerobically from mannitol, production of phosphatase and examination for arginine dihydrolase activity has been proposed by Kocur (personal communication) for the differentiation of the three recognised *Staphylococcus* species. This scheme gave very satisfactory results in an investigation of urinary strains by Digranes and Oeding (1975), the biochemical classification corresponding closely to type of teichoic acid and Novobiocin sensitivity. A somewhat different scheme was recently (1976) recommended by "The ICSB Subcommittee on the Taxonomy of Staphylococci and Micrococci". Strains of staphylococci not identified by these schemes are probably of minor medical interest. If required such strains should be classified by the criteria used by Kloos and Schleifer (1975b) for the description of seven new species.

In animals, variants of *S. aureus* have developed as a result of adaptation to the host conditions and several such biotypes or ecotypes have been described (Meyer, 1967; Hájek and Maršálek, 1971). A subdivision of coagulase-positive staphylococci from man and from different animal species is of practical value for veterinary and food microbiologists.

## III. CELL WALL

The cell walls of staphylococci contain two major structures: The teichoic acid and the peptidoglycan. In addition, a variety of minor protein and carbohydrate components possessing antigenic activity are present. The cell walls seem usually to be surrounded by a loose layer of material and certain strains possess a true capsule.

### A. Teichoic acids

Baddiley and associates demonstrated that the walls of *S. aureus* contain a ribitol teichoic acid with N-acetylglucosamine substituents (Armstrong *et al.*, 1958; Baddiley *et al.*, 1961). Both α- and β-anomers of the sugar may be present, and their proportions in the teichoic acids differ in different strains (Nathenson *et al.*, 1966). The N-acetylglucosaminyl ribitol teichoic acid has been considered specific for the walls of *S. aureus* (Archibald, 1972).

The antigenic activity of the *S. aureus* wall teichoic acid was first demonstrated by Julianelle and Wieghard (1934, 1935a, b) and rediscovered independently by Juergens *et al.* (1960), Haukenes *et al.* (1961a, b) and Morse (1962). The N-acetylglucosaminyl ribitol unit was found to be the serological determinant, the specificity depending on the configuration of the glucosaminyl substituents (Juergens *et al.*, 1960; Sanderson *et al.*, 1961; Haukenes *et al.*, 1961a, b; Morse, 1962). In agar gel diffusion α-linked N-acetylglucosamine in the teichoic acid gave one specific precipitation line and β-linked N-acetylglucosamine another line (Haukenes *et al.*, 1961a, b; Haukenes, 1962b; Davison *et al.*, 1964; Hofstad, 1965b). According to serological investigations (Marandon and Oeding, 1967), the β-linkage is more common than the α-linkage in naturally occurring strains of *S. aureus*. Both purified polysaccharide (see p. 137) and teichoic acid behave as haptens on immunisation of rabbits (Haukenes, 1962b), whole bacteria therefore being used for the production of antisera.

The species-specificity of the N-acetylglucosaminyl ribitol teichoic acid for *S. aureus* was later shown to be relative. Phage 187-susceptible strains have been shown to contain N-acetyl galactosamine instead of N-acetyl glucosamine in their teichoic acid, giving it a new serologic specificity (Karakawa and Kane, 1971; Oeding, 1974; Galinski and Lipinska, 1974). By agar precipitation of coagulase-positive staphylococci isolated from various animal species and classified into biotypes according to Hájek and Maršálek (1971), several polysaccharides have been demonstrated (Table I). Bovine, hare, typical poultry and typical swine strains regularly contain β-N-acetylglucosaminyl ribitol teichoic acid (Oeding, 1973; Devriese and Oeding, 1976) whereas equine strains contain α-N-acetylglucosaminyl ribitol teichoic acid. Dog strains contain an unusual teichoic acid (poly P) with repeating units of glycerol, glucosamine and α-glucose (Oeding, 1973; Endresen *et al.*, 1974; Endresen and Grov, 1976) and pigeon and mink strains a β-glucosaminyl glycerol teichoic acid (Oeding, 1973; Endresen *et al.*, 1974) apparently identical to poly C of *S. saprophyticus* (Losnegard and Oeding 1963a, b; Johnsen *et al.*, 1975a, b). Poly C was also detected in biochemically atypical poultry and swine strains (*S. hyicus*) together with an unidentified polysaccharide (poly A1) (Devriese and Oeding, 1975).

The walls of coagulase-negative staphylococci have been shown to contain a glycerol teichoic acid with glucosyl substituents occurring in an α- or β-position (Davison and Baddiley, 1963, 1964; Losnegard and Oeding, 1963a; Morse, 1963). This antigen was also described by Julianelle and Wieghard (1934, 1935a). Both the α- and the β-linked glucose are serologically active and produce specific precipitation lines in agar (Morse, 1963; Losnegard and Oeding, 1963b; Davison *et al.*, 1964; Oeding *et al.*, 1967). The α-type is much more common than the β-type in naturally occurring coagulase-negative staphylococci (Morse, 1963; Losnegard and Oeding, 1963b; Aasen and Oeding, 1971). The glucosyl glycerol teichoic acid seems to be characteristic of *S. epidermidis sensu stricto*, corresponding to Baird-Parker's biotype 1 (1963) (Digranes and Oeding, 1975; Schleifer and Kloos, 1975a).

*S. saprophyticus* contains two teichoic acids in its walls. One is chemically and serologically identical to the $\beta$-N-acetylglucosaminyl ribitol teichoic acid present in *S. aureus* walls, while the other is a $\beta$-N-acetylglucosaminyl glycerol teichoic acid producing another precipitation line in agar (Losnegard and Oeding, 1963a, b; Johnsen *et al.*, 1975a, b). Still other teichoic acids are present in staphylococci classified in Baird-Parker's biotypes 2 to 5 (1963) (Digranes and Oeding, 1975), and in the new species proposed by Schleifer and Kloos (1975a) and Kloos and Schleifer (1975a).

## B. Peptidoglycan

Most strains of staphylococci contain a peptidoglycan crosslinked by penta- or hexaglycine bridges (Schleifer and Kandler, 1972). Usually between 3 and 6 moles of glycine residues are found per mole of peptide subunit and the high content of glycine is typical for the peptidoglycans of staphylococci. In *S. aureus* a small amount of glycine is sometimes replaced by L-serine, whereas *S. epidermidis* contains a significantly higher amount of L-serine residues. Several types of staphylococcal peptidoglycans have been demonstrated chemically (Schleifer and Kandler, 1972; Schleifer and Kloos, 1975a; Kloos and Schleifer, 1975a). The peptidoglycan has been shown to possess serological properties and antibodies directed against the glycan and the peptide subunits have been demonstrated (Oeding *et al.*, 1964; Karakawa *et al.*, 1968; Grov 1969a, b; Helgeland *et al.*, 1973). The peptide moiety of the *S. aureus* peptidoglycan possesses several antigenic specificities when examined by indirect haemagglutination (Grov and Oeding, 1971; Helgeland *et al.*, 1973).

## C. Potein A

Protein A (Verwey, 1940; Jensen, 1958) is a wall constituent of *S. aureus* containing 14–16% nitrogen and several amino-acids (Yoshida *et al.*, 1963; Grov *et al.*, 1964; Löfkvist and Sjöquist, 1964). A characteristic property is its reaction with normal sera (Jensen, 1958) through the Fc fragment of IgG (Forsgren and Sjöquist, 1966). Protein A is conveniently demonstrated by agglutination of red cells coated with IgG or by agar precipitation. While haemagglutination is more sensitive, the precipitation reaction is preferred for the simultaneous screening of polysaccharides and protein A. Using agar precipitation with normal human serum, protein A has been shown to be regularly present in human strains of *S. aureus* (Oeding and Haukenes, 1963), less frequently in bovine strains (Marandon and Oeding, 1967; Oeding *et al.*, 1971) and infrequently or not at all in coagulase-positive staphylococci of other biotypes (Oeding *et al.*, 1972; Oeding *et al.*, 1973; Devriese and Oeding, 1975, 1976). Protein A is not present in *S. epidermidis* (Aasen and Oeding, 1971).

## D. Type-specific agglutinogens

Staphylococci contain a large number of agglutinogens which have been located in the cell wall (Grov and Rude, 1967). Cross-absorption experiments and the slide agglutination technique have revealed that the majority are species-specific, whereas some are shared. Serological typing systems are available for *S. aureus* (Cowan, 1939; Mercier *et al.*, 1950; Oeding, 1952a, b), but not yet for *S. epidermidis*, although this organism also contains species-specific agglutinogens (Aasen and Oeding, 1971).

Little is known of the chemical nature of the type agglutinogens. The results of heating and treatment with proteolytic enzymes indicate that most agglutinogens are proteins, whereas some are carbohydrates (Pillet *et al.*, 1955; Haukenes and Oeding, 1960; Haukenes, 1967). Isolation of these antigens, which occur in very small amounts in the cell wall, is difficult. At least two agglutinogens produce precipitation lines in agar which facilitate tracing during purification. The isolation of the *n* agglutinogen resulted in one major polypeptide component responsible for agglutination and another minor carbohydrate component responsible for precipitation (Grov *et al.*, 1966). Work in progress at this Institute indicates that agglutinogen $h_1$ is a protein containing 10–12 amino-acids.

## E. Surface material

Blocking of agglutination has been found to be a regular phenomenon in *S. aureus* strains (Oeding, 1953). Autoclaving and treatment of the bacteria with trypsin made the blocked antigens agglutinable (Oeding, 1953). Blocking was also found to be a function of duration and temperature of incubation of the bacteria (Oeding, 1957) and of the composition of the culture medium (Haukenes, 1962b). Adsorption of phages to the receptors on the *S. aureus* surface has been shown to be inhibited by protein A (Forsgren and Nordström, 1974). The blocking substances may be integrated parts of the outer cell surface but a loose layer surrounding the cell may also be significant.

True capsules containing polysaccharide antigen have been demonstrated in mucoid *S. aureus* strains (Wiley, 1972). Furthermore, an extracellular muco-polysaccharide having antigenic properties is produced under certain growth conditions (Ekstedt, 1974). Although these substances are interesting in relation to pathogenicity, they will not be considered further here.

## IV. IDENTIFICATION OF SPECIES AND BIOTYPES BY PRECIPITATION

### A. Teichoic acids

The type of teichoic acid is a fundamental criterion in the taxonomy and classification of the *Micrococcaceae*. Chemical determination is too complicated for diagnostic and epidemiological purposes. The serological specificity of the teichoic acids has been demonstrated by inhibition of cell wall agglutination (Juergens *et al.*, 1960), haemagglutination, precipitation inhibition (Morse, 1962) and tube test and agar precipitation combined with inhibition reactions (Haukenes *et al.*, 1961b; Haukenes, 1962c; Davison *et al.*, 1964; Hofstad, 1965b). The method of choice for epidemiological purposes is precipitation in agar gel using reference systems (Oeding, 1973).

In the following these substances are – for historical and practical reasons – called polysaccharides. Carbohydrate A and B were used by Julianelle and Wieghard (1934, 1935a, b) to designate the species-specific antigens of *S. aureus* and *S. epidermidis* respectively. These substances were shown to correspond to the teichoic acids of the two species (Haukenes *et al.*, 1961a; Davison *et al.*, 1964; Oeding *et al.*, 1967) and have in the papers from the Institute been called polysaccharide (poly) A and B (Haukenes *et al.*, 1961b; Losnegard and Oeding, 1963a). Similar staphylococcal wall substances described later have been given corresponding designations (Oeding, 1973). This may also be the most correct since pure teichoic acids are not used in the routine serology procedure. Furthermore, the designation allows the inclusion of polysaccharides whose structures have not yet been determined, in addition to being practical for use in an abbreviated form.

*Designations:*

Poly $A_\alpha$ = $\alpha$-N-acetylglucosaminyl ribitol teichoic acid
Poly $A_\beta$ = $\beta$-N-acetylglucosaminyl ribitol teichoic acid
Poly $B_\alpha$ = $\alpha$-glucosyl glycerol teichoic acid
Poly $B_\beta$ = $\beta$-glucosyl glycerol teichoic acid
Poly C = $\beta$-N-acetylglucosaminyl glycerol teichoic acid
Poly 187 = N-acetylgalactosaminyl ribitol teichoic acid
Poly P = unusual teichoic acid with repeating units of glycerol, glucosamine and $\alpha$-glucose
Poly $A_1$ = N-acetylglucosaminyl glycerol teichoic acid.

### 1. *Agar precipitation test*

A modified Ouchterlony double diffusion in agar technique is used (Oeding, 1973). Plastic Petri dishes with a diameter of approximately

85 mm and a flat, unscratched bottom are poured with 12 ml quantities of 1·2% (w/v) agar (Noble, Difco) in distilled water containing 0·85% (w/v) NaCl. Buffering may result in unclear agar and should be avoided. The agar is sterilised by steaming at 100°C. Circular wells are cut with a cork borer using a paper model clearly visible through the agar layer. Systems are arranged consisting of 6 peripheral (No. 1–6) and 1 central wells (No. 7) 3 mm in diameter and 4 mm apart. Small amounts of reagents are needed to fill the wells. Up to 6 systems may be arranged in one agar plate (Fig. 1). The central well is filled with undiluted antiserum, the peripheral wells No. 2, 3, 5 and 6 with the bacteria to be tested and wells No. 1 and 4 with

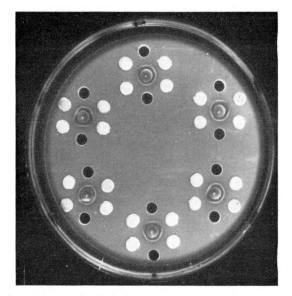

FIG. 1. Testing presence of poly $A_\beta$ in 24 strains of S. *aureus* by agar precipitation. Antiserum to S. *aureus* Wood 46 in central well, poly $A_\beta$ in wells 1 and 4, bacteria in wells 2, 3, 5 and 6. Poly $A_\beta$ in all strains except six.

the polysaccharide reference. In the poly $A_\beta C$ system, polysaccharide $A_\beta$ is included as an additional reference, replacing bacteria in well No. 2. Antiserum and polysaccharide are added with a thin Pasteur pipette by allowing the fluid to run down the side of the well. Bacteria are removed from an agar slant and added to the wells using a small loop.

The plates are kept at 4°C and the reactions read under a lamp daily for at least 2 days. As a rule the precipitation lines are most sharp and the determination or identity or non-identity easiest after 1 day (Fig. 2). After 2 days weak reactions may have developed and identity reactions which were questionable after 1 day may be ascertained. The lines which are strong after 1 day often become diffuse after 2 days. Reading after 3 days seldom gives additional information. A reaction of identity with the standard polysaccharide is required for identification.

## 2. *Bacteria*

The strains to be tested are grown on nutrient agar at 37°C overnight (Oxoid agar 1·3% (w/v), proteose peptone 1·5% (w/v), yeast extract 0·5% (w/v), liver digest 0·25% (w/v), pH 7·3–7·4). The moist surface growth is removed with a small loop and added to the wells as described above.

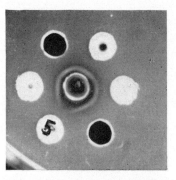

FIG. 2. Detail of Fig. 1. Poly $A_\beta$ in three strains, not in No. 5.

## 3. *Antisera*

Lyophilised bacteria of the reference strains are tested for purity on blood agar plates from which nutrient agar slants are inoculated. After 18 h incubation at 37°C, the growth is suspended in a 1% (w/v) solution of formalin (40%) and kept at 37°C for 24 h (Oeding, 1957). After centrifugation and sterility control on blood agar plates, a suspension in saline of approximately $5 \times 10^9$ bacteria per ml is made. New Zealand white rabbits of the Institute's breed are injected intravenously on 3 successive days with an interval of 5 days to the next series. Doses of 0·1, 0·2, and 0·4 ml are given in the first series, 0·4, 0·6, and 0·8 ml in the second, and 0·8, 1·0, and 1·0 in the third. Usually after 3 series and bleeding 5 days after the last injection, satisfactory antisera are obtained. The sera are inactivated at 56°C for 30 min. and 5 ml portions of serum, to which 2 drops of a 1% (w/v) solution of merthiolate have been added, are distributed in small glass tubes with rubber stoppers and frozen.

To be considered satisfactory, a serum should produce a strong precipitation line with the homologous reference polysaccharide after 1 day.

A good serum will keep for years. Occasionally, satisfactory sera may be difficult to obtain. Weak precipitation lines or no line whatsoever may be the result. Continuation of immunisation may render the serum usable, but usually the serum will not be satisfactory and new rabbits should be selected for immunisation. It is advisable to immunise 2 or 3 rabbits at the same time with the same vaccine, since variation in individual rabbit response is probably the main cause of the difficulty in obtaining satisfactory antisera. Also, the amount of polysaccharide in the reference strain and its antigenicity may play a role. Attempts to overcome these difficulties have been made using other immunisation techniques and other strains but a final solution to the problem has not been found.

## 4. *Polysaccharides*

The procedures for the isolation of polysaccharides are essentially as described by Haukenes (1962a). Formerly bacteria were harvested from 18 h nutrient agar plates 14 cm in diameter by scraping the surface. In order to save media and time, fluid culture is now preferred. One hundred millilitres of nutrient broth (Oxoid) in a 200 ml bottle are inoculated with the bacteria, incubated at 37°C in an incubator shaker (New Brunswick Scientific Co., Inc., New Brunswick, N.J., U.S.A.) for 6 h at 150 rev/min, transferred to 1000 ml nutrient broth in a 2000 ml bottle and further incubated for 18 h, after which the bacteria are spun down. The wet bacteria are extracted with a 1/15 M phosphate buffer, pH 6·5 for 24 h at 37°C and the bacteria are removed by centrifugation. The extraction is usually repeated thrice, sometimes 4 to 6 times, to get maximum yield. For the first extraction 10 ml of buffer per g bacteria (wet weight) are used, for the other extractions 5 ml. The supernatants are collected and the pH adjusted to 4·2 with 1 N HCl to remove proteins. After 24 h at 4°C the precipitate is removed by centrifugation. The pH of the supernatant is raised to 5·2 with 1 N NaOH and the polysaccharide precipitated with 5 volumes of ethanol at 4°C for 24 h. The precipitate is collected by centrifugation, dissolved in distilled water and lyophilised.

The crude polysaccharide is purified by ion exchange chromatography and gel filtration. DEAE (diethylaminoethyl)-cellulose, 100–230 mesh, is equilibrated first with 0·2 M phosphate buffer, pH 7·4 and then with 0·02 M, and packed in columns of approximately 2·5 × 50 cm. Thereafter approximately 400 mg crude polysaccharide is applied. Elution is performed with 0·02M phosphate buffer, pH 7·4 (approximately 200 ml) followed by a gradient system (total volume 1000 ml) from 0 to 1 M KCl. The flow rate is approximately 40 ml/h. Fractions are collected and the optical density read at 280 nm. The fractions found to be positive with homologous antiserum in the ring test are pooled, dialysed and lyophilised. Further purification is performed by gel filtration on Sephadex G-75 columns (approximately $2.5 \times 95$ cm) using a $0.1$M Tris-HCl buffer, pH $8.0$, with $0.5$M KCl. The final material contains a small amount of peptidoglycan fragments which are of no consequence for the agar precipitation test.

The purified polysaccharides are lyophilised and used in the agar precipitation tests in a saline dilution producing a strong and suitably situated line with the homologous antisera. Usually a $1:20$ dilution of a 1 mg/ml solution of the polysaccharide will give a good standard.

### 5. *Reference systems*

Reference systems regularly used:

*Poly A$_\alpha$* and antiserum produced from *S. aureus* 263 (Hofstad, 1965a). The system produces a poly A$_\beta$ line as well.

*Poly A$_\beta$* and antiserum produced from *S. aureus* Wood 46 (Haukenes, 1962a).

*Poly B$_\alpha$* and antiserum produced from *S. epidermidis* 1254 (Losnegard and Oeding, 1963b).

*Poly B$_\beta$* and antiserum produced from *S. epidermidis* T2 (Oeding *et al.*, 1967).

*Poly C* and antiserum produced from *S. saprophyticus* 3519 (Losnegard and Oeding, 1963b).

Additional reference systems:

*Poly* 187 and antiserum produced from *S. aureus* 2243 (Oeding, 1974).

*Poly P* and antiserum produced from *S. aureus* canine biotype Z14 (Oeding, 1973).

## B. Protein A

Protein A is present especially in the human biotype of *S. aureus* and never in coagulase-negative staphylococci. The demonstration of protein A may therefore be of some diagnostic interest. Protein A is produced from *S. aureus* Cowan I (NCTC 8530) which has a high content of this antigen.

Wet bacteria are obtained either from nutrient agar plates or from nutrient broth shake cultures. Extraction and purification are performed essentially as described by Grov *et al.* (1964). The bacteria are suspended in a 1/15M phosphate buffer, pH $5.9$ (approximately 10 ml per g bacteria) and boiled for 1 h in a water bath with constant stirring. After cooling, the bacteria are spun down and the extraction repeated once, sometimes twice, with 5 ml buffer per g to get maximum yield. The pH of the supernatants is adjusted with 1 N HCl to 3, the supernatants are kept in the ice box overnight and the precipitates are spun down. Precipitates showing a satisfactory activity on agar precipitation are collected, dissolved in a small volume of distilled water by the dropwise addition of the necessary quantity of 1N NaOH, dialysed against a continuous flow of tap water for 24 h and finally against distilled water for 24 h.

Ion exchange chromatography is used for further purification of the antigenic material. DEAE (diethylaminoethyl)-cellulose, 100–230 mesh, is suspended in 1M ammonium formate, pH 6·8, and applied to the column. First 100–200 ml of 1M ammonium formate is run through the column and then distilled water until neutrality is reached. The antigenic material is applied in as high a concentration as possible. Elution is performed first with approximately 150 ml of distilled water, then with an ammonium formate (pH 6·8) gradient from 0 to 1M (total volume 1000 ml). The fractions are examined spectrophotometrically at 280 nm, by ring test precipitation, and by agar precipitation. Fractions showing protein A are collected, concentrated *in vacuo*, dialysed against tap water and distilled water as above, and lyophilised. Then the column fractionation is repeated once. The final protein A preparation contains small amounts of protein B (Grov *et al.*, 1964) and peptidoglycan, these substances being of no consequence in the agar precipitation test.

In the agar precipitation test the central well (No. 7) is filled with a pool of 10 normal human sera, wells 1 and 4 with protein A and wells 2, 3, 5. and 6 with the bacteria to be tested. The concentration of protein A producing a strong and sharp precipitation line is determined. As a rule a 1 mg/ml solution in saline is used. The plates are read after 1 day's incubation at 37°C and checked after a second day at room temperature. The majority of strains after 1 day present a strong line fusing with the standard protein A line. After 2 days lines may be visible in strains producing little protein A while the old lines have now become diffuse.

## C. Field examinations

### 1. S. aureus *strains from man and his environment*

Phage typing and/or serological typing usually provide satisfactory epidemiological information for coagulase-positive strains of human origin. Strains not containing the typical wall teichoic acid (poly A) but another teichoic acid, most often poly 187, can be detected by agar precipitation. However, such strains are rather uncommon and the strains containing poly 187 can be detected by phage-typing (lysed by phage 187 only) and by serological typing (agglutinogen *k*). Non-typability by phage is increasing in frequency and may present epidemiological problems. If the strain responsible for an epidemic outbreak has a defined phage pattern the non-typable strains are of minor interest. But should an unusually high percentage of the isolated strains be non-typable, and the epidemic strain also be non-typable or have an ill-defined phage pattern, additional characterisation is desirable. This is also true when the epidemic strain has a very common phage pattern. Although human biotype strains as a rule contain both poly A and protein A, there are nevertheless qualitative and quantitative variations. Differences in the content of poly $A_\alpha$, poly $A_\beta$ and protein A may contribute to the subdivision of the isolated strains. However, examination of wall precipitinogens is of minor interest in the epidemiological characterisation of *S. aureus* strains of human origin.

## 2. S. aureus *strains from animals and their environment*

Cross colonisation between different animal species and between man and animals in animal houses and research stations is well known (Live, 1972a). When outbreaks of staphylococcal infection occur in animals, the causative strains are regularly of the adapted biotype. However, epidemiological surveys often reveal a certain number of strains which are different, some being of another biotype, for example of the human one, others having intermediary characteristics. The type of wall teichoic acid is a characteristic property of strains from different animals (Table I).

In this situation, serological determination of the type of teichoic acid and the presence of protein A is a valuable supplement to the biochemical subdivision of the strains.

Two examples illustrate the contribution to classification given by serological determination of teichoic acids. In Table II the biochemical

TABLE I

**Presence of wall polysaccharides and protein A in *S. aureus* strains from animal species**

|            | Poly $A_\alpha$ | Poly $A_\beta$ | Poly C | Poly P | Protein A |
|------------|:---:|:---:|:---:|:---:|:---:|
| Man[1]     | + | + | − | − | + |
| Cattle[1]  | − | + | − | − | ±[†] |
| Sheep[3]   | − | + | − | − | ±[†] |
| Poultry[2] | − | + | − | − | ±[†] |
| Swine[1,2] | − | + | − | − | − |
| Horse[1]   | + | − | − | − | − |
| Dog[1]     | − | − | − | + | − |
| Pigeon[1]  | − | − | + | − | − |
| Mink[1]    | − | − | + | − | − |

[†] Some strains contain protein A
[1] Oeding (1973)
[2] Devriese and Oeding (1976)
[3] Oeding et al. (1976)

classification showed that 10 of the 75 *S. aureus* strains isolated from dogs corresponded to the human biotype (Hájek and Maršálek, 1969), apparently having been transmitted from human contacts. The demonstration of poly $A_\alpha$ and $A_\beta$ in these strains substantiated the classification. The canine biotype strains did not contain poly A but poly P.

TABLE II

**S. aureus strains isolated from the nares of dogs** (data from Oeding, 1973)

| Biotype | | Polysaccharide | | |
|---------|-----|-----|-----|-----|
| | | $A_\alpha$ | $A_\beta$ | P |
| Canine | 63 | 0 | 0 | 52 |
| Human | 10 | 6 | 9 | 0 |
| Intermed. | 2 | 0 | 0 | 1 |

TABLE III

**Coagulase-positive staphylococci isolated from poultry and swine** (data from Devriese and Oeding, 1975, 1976)

| | | Polysaccharide | | |
|---|-----|-----|-----|-----|
| | | $A_\beta$ | $A_1$ | $A_1C$ |
| Poultry and swine, biotype B | 203 | 193 | 0 | 0 |
| Poultry and swine, atypical | 63 | 0 | 23 | 40 |

In Table III the precipitation reactions corresponded very well with the results of the biochemical classification. Ninety-five per cent of the poultry and swine strains belonging to biotype B contained poly $A_\beta$, whereas none of the 63 atypical strains did so. These strains all contained an atypical polysaccharide ($A_1$) and the majority of the strains poly C as well.

## 3. S. epidermidis *strains from man*

Coagulase-negative staphylococci are not infrequently isolated from the blood in cases of septicaemia and subacute bacterial endocarditis (Smith *et al.*, 1958; Quinn *et al.*, 1965; Shulman and Nahmias, 1972). The causative organism seems as a rule to belong to *S. epidermidis* biotype 1 (Baird-Parker, 1974a). These strains contain poly B and according to our experience the glucose is nearly always $\alpha$-linked. Thus of 21 coagulase-negative staphylococci isolated from blood culture of patients with mostly post-cardiac surgery endocarditis, 20 strains contained poly $B_\alpha$ when examined by us (Smith and Farkas-Himsley, 1969). The demonstration of this teichoic acid excludes an atypical coagulase-negative *S. aureus* strain.

When repeated *S. epidermidis* infections occur in a ward an epidemiological survey may be required. Many different strains of coagulase-negative staphylococci as well as micrococci may be isolated. Determination of the techoic acid by means of agar precipitation will then be a rapid and reliable method for the identification of biotype 1 strains.

### 4. S. saprophyticus *strains from man*

Members of the family *Micrococcaceae* are responsible for approximately 5% of urinary tract infections (Digranes and Oeding, 1975). *S. saprophyticus* probably is the most common, being isolated nearly exclusively from female patients seen in general practice, while *S. aureus* and *S. epidermidis* are found particularly in hospitalised male patients (Roberts, 1967; Maskell, 1974; Digranes and Oeding, 1975). The demonstration of *S. saprophyticus* in urinary tract infections of hospitalised patients thus contradicts hospital transmittance. Epidemiological measures should therefore, if required, be directed against *S. aureus* and *S. epidermidis* infections. The two *S. saprophyticus* teichoic acids, *viz.* poly A and poly C are easily separated from poly B of *S. epidermidis* biotype 1 by agar precipitation, thus permitting an identification of the two species and a separation from other staphylococcal biotypes and micrococci.

### D. Advantages and problems of the procedure

The test is rapid and easy to perform, small amounts of reagents are required and the results are usually obtained after 1 day. Positive results are reliable providing that sufficient experience with the agar precipitation technique and satisfactory control systems exist. Even if the investigator has long experience with the method, controls should never be omitted. The test provides information which may be of value for classification and epidemiological evaluation of staphylococcal infections.

The precipitation reaction is, however, not very sensitive and may fail to recognise an antigen present in small amounts. Thus the presence of poly A in strains negative on direct precipitation may be revealed by absorption experiments (Haukenes, 1962b). A weak or negative precipitation reaction is a stable characteristic and not a result of technical variation. Such reactions therefore contribute to the epidemiological characterisation of strains. A more serious practical problem is the occasional failure in obtaining satisfactory antisera even after repeated immunisations. As a rule, however, serum production presents no problem and the sera keep well for several years.

## V. SUBDIVISION OF *S. AUREUS* BY AGGLUTINATION

### A. Historical

Cowan (1939) was the first to report a method allowing serological differentiation of *S. aureus* strains. An important technical detail was the performing of the agglutination on slides rather than in tubes. This procedure gave stable suspensions of the bacteria and has been adopted by later investigators. Boiled bacteria were used for the production of antisera, for absorption, and for agglutination. This was done to reduce cross-reactions, thereby facilitating serotyping. Absorbed and unabsorbed antisera produced against three *S. aureus* strains allowed the differentiation into three types and a miscellaneous group. The strains are today known as the Cowan I, II and III strains. To achieve better differentiation new strains were included and the number of typing sera were increased to 13 (Gillespie *et al.*, 1939; Christie and Keogh, 1940; Hobbs, 1948).

Although Cowan's method was used with success in some field investigations (Gillespie *et al.*, 1939; Elliot *et al.*, 1941; Hobbs, 1948) it had obvious deficiencies and was not widely adopted. Important objections were that the typing sera were polyvalent and did not permit a satisfactory type differentiation, and that many strains were non-typable. It was noticed that strains of *S. aureus* usually contain several agglutinogens and Hobbs (1948) suggested that patterns of reactions might be useful for serological identification.

Attempts to improve Cowan's method were made by Pereira (1961) and by Pillet and associates who since 1950 (Mercier *et al.*, 1950) have published a series of reports. Pillet also used autoclaved bacteria for immunisation and absorption. The composition of the culture medium was found to affect the results of agglutination (Pillet *et al.*, 1951). The strains to be tested were grown in a broth medium, the bacteria were spun down and then agglutinated on a large glass plate in a mechanical agitator. The procedures have been described in detail by Pillet *et al.* (1967). First the number of type strains was extended to 18 (Pillet *et al.*, 1961; Pillet *et al.*, 1966) and later reduced to 14 (Pillet *et al.*, 1967). The aim was to obtain relatively defined antisera which reacted with the homologous strain only. Such sera were not obtained because several sera produced one dominant reaction as well as cross-reactions.

The methods of Cowan and Pillet registered the heat-stable agglutinogens only. Andersen (1943) and Oeding (1952a) showed that this pro-

6

cedure was not usable for typing strains responsible for outbreaks of pemphigus neonatorum and mastitis. When antibodies against heat-labile agglutinogens were included, the strains became typable.

Based on these experiences, Oeding (1952b, 1960) developed a new method for serotyping *S. aureus*. This method is different in principle from those earlier described. After absorption, the immune sera are specific not for a strain but for an agglutinogen. The ten type-specific agglutinogens first detected by cross-absorptions were designated by small letters, and the agglutinations of a strain in the set of absorbed sera, called factor sera, were recorded as an antigenic pattern. This allowed a better differentiation of strains than a system which is bound to a limited number of types. In Oeding's method formalin-killed bacteria were used for the production of rabbit immune sera and live bacteria were used for absorption and for agglutination on slides.

The condition of the bacteria was shown to be decisive for the outcome of agglutination. By using live bacteria, heat-labile agglutinogens could be detected as well. Further investigations, however, revealed that live bacteria from cultures incubated for 18 h often contained antigens and other material on their surface which blocked the agglutination of other antigens (Oeding, 1953, 1957). This effect being significant in practical work, was overcome by using bacteria from 5 h cultures (Oeding, 1957) or mannitol-salt agar cultures (Haukenes, 1962b) for agglutination. These studies also gave an indication of the distribution of the antigens on the cell surface and of their chemical nature.

The discovery of the blocking effect upon agglutination permitted a more detailed study of the antibodies present in the factor sera. The three original Cowan sera were shown by cross-absorption to contain a complex of antibodies (Haukenes, 1964c), and also several of the Oeding factor sera contained more than one antibody (Haukenes and Oeding, 1960; Haukenes, 1967). The original Oeding method included his own strains in addition to three strains isolated by Andersen (1943). Several new strains were now included, among them the three Cowan strains and strain Wood 46. The new strains helped to obtain strong, monospecific factor sera and to identify the different antibodies present in the antisera. This resulted in the demonstration of a number of new agglutinations. The results were reported by Haukenes in a series of papers and have been reviewed by Haukenes (1967). Additional agglutinogens were detected by Hofstad (Hofstad and Oeding, 1962; Hofstad 1964a), including agglutinogen 263–1, which correlated closely with the phage 80/81 complex, and by Grün (1958a, 1961).

This work resulted in a revised scheme containing 13 factor sera which were sufficiently strong and defined to be suitable for routine typing (see below). Several new agglutinogens were not included because the antibody titres were low and more suitable type strains were not available. To increase the number of factor sera further would also create practical problems. The revised scheme, in addition to recording a larger number of agglutinogens, also increased the typability.

Both Pillet's and Oeding's method have been used in a series of field examinations and have provided valuable epidemiological information. Oeding's method has, however, been increasingly preferred and is today the prevailing method for serotyping *S. aureus*. A few comparative typings with the two methods have been performed in order to provide the basis for one single method (Pillet *et al.*, 1967; Galinski and Krynski, 1970; Modjadedy and Fleurette, 1974). Fleurette and Modjadedy (1976) recently stated that a better antigenic analysis is obtained by Oeding's method than by Pillet's. They suggest that Oeding's method should be used as the standard procedure, but that a specified fluid medium be introduced for the growth of the strains, and Pillet's modification be used for the agglutination reaction. This seems to be an important step towards a standardised method which can be internationally acceptable. Further investigations are, however, needed and until a final decision can be reached the present Oeding method is recommended for the typing of *S. aureus*.

## B. Oeding's method

### 1. *Strains*

Twenty strains of *S. aureus*, all positive in the tube coagulase test with rabbit plasma and all of human origin are used for the production and absorption of antisera, and for the checking of factor sera. A survey of the strains is given in Table IV.

The strains have been preserved freeze-dried since their isolation or acceptance. On request, numerous sets of the freeze-dried strains have been sent to institutes all over the world. Vaccine for immunisation is produced from freeze-dried bacteria. During the period of immunisation, absorption and testing of non-absorbed and absorbed antisera, the strains are kept on nutrient agar (Oxoid), preferably in deep culture in screw-cap bottles, or alternatively on slants.

### 2. *Antisera*

Pre-immune rabbit antibodies agglutinating staphylococci in low titres are regularly found, varying from one animal house to another, and depending on the bacterial strains examined. Several different agglutinins are involved. To avoid false positive reactions, serum should be taken

## TABLE IV

### *S. aureus* strains used in the typing system

| No. | Origin | Serological reference |
| --- | --- | --- |
| 28 | Mastitis, Dpt. Microbiol., Gade Inst., Bergen | Oeding (1952b) |
| 1503 | Mastitis, Dpt. Microbiol., Gade Inst., Bergen | Oeding (1952b) |
| 3647 | Panaritium, Dpt. Microbiol., Gade Inst., Bergen | Oeding (1952b) |
| 263 | Carbuncle, Dept. Microbiol., Gade Inst., Bergen | Hofstad (1964a) |
| 830 | Food Poisoning, Dpt. Microbiol., Gade Inst., Bergen | Haukenes 1963a) |
| 137 | Nose/throat carrier, Dpt. Microbiol., Gade Inst., Bergen | Hofstad (1964a) |
| 2095 | Nose/throat carrier, Dpt. Microbiol., Gade Inst., Bergen | Oeding (1952b) |
| 2253 | Nose/throat carrier, Dpt. Microbiol., Gade Inst., Bergen | Oeding (1952b) |
| 3189 | Nose/throat carrier, Dpt. Microbiol., Gade Inst., Bergen | Oeding (1952b) |
| 17A | Pemphigus neonat., State Serum Inst., Copenhagen | Oeding (1952b) |
| F21 | Pemphigus neonat., State Serum Inst., Copenhagen | Oeding (1952b) |
| S365 | Nose carrier, State Serum Inst., Copenhagen | Oeding (1952b) |
| C.I | NCTC 8530, Publ. Hlth Lab. Serv., Colindale | Haukenes and Oeding (1960), Haukenes (1964c) |
| C.II | NCTC 8531, Publ. Hlth Lab. Serv., Colindale | Haukenes and Oeding (1960), Haukenes (1964c) |
| C.III | NCTC 8532, Publ. Hlth Lab. Serv., Colindale | Haukenes and Oeding (1960), Haukenes (1964c) |
| W.46 | Publ. Hlth Lab. Serv., Colindale | Haukenes and Oeding (1960), Haukenes (1964c) |
| 5687 | Publ. Hlth Lab. Serv., Colindale | Haukenes (1964a) |
| 6376 | Food Poisoning, Publ. Hlth Lab. Serv., Colindale | Haukenes 1963b) |
| 670 | Hyg.–Inst. Bautzen | Haukenes (1964a) |
| 1015 | Hyg.–Inst. Bautzen | Haukenes (1963b) |

C.: Cowan strain
W.: Wood strain

prior to immunisation and tested by slide agglutination, in a $1:10$ saline dilution, with all the 20 type strains (Haukenes, 1967). Thereafter the capacity to remove eventual pre-immune agglutinins is tested with the strain or strains later to be used for absorption of the antiserum. If the pre-immune agglutinins are removed, the rabbit can be used for the production of that particular factor serum. If the pre-immune agglutinins are not removed, the rabbit should be rejected.

Table V illustrates the procedure. Five strains are agglutinated by the pre-immune serum diluted $1:10$. The rabbit was intended for the preparation of factor serum $a_4$, which is made from antiserum to strain 3647

TABLE V

**Agglutination of *S. aureus* type strains in a pre-immune serum before and after absorption**
(Haukenes, 1967)

| Strain | Agglutination in pre-immune serum ($1:10$) | |
|---|---|---|
| | Not absorbed | Absorbed with strain F21 |
| 1503 | + | − |
| 28 | + + | − |
| 17A | + + | − |
| C.I | + + + | − |
| 830 | + | − |
| Other type strains | − | − |

by absorption with F21, 1503, 17A and 3189 bacteria. Strain F21 removed all pre-immune agglutinins and the rabbit could therefore be used for the intended purpose.

When suitable rabbits have been selected immunisation is started. Preparation of vaccine and the immunisation schedule have been described on page 136. Antisera are produced against *S. aureus* strains 1503, 3647, 263, 17A, F21, S365, C.II, and 5687 (Table VII). Rabbits vary considerably in their antibody response to staphylococcal agglutinogens. Some rabbits are generally poor antibody producers, others produce one antibody abundantly, but little of another. This is illustrated in Table VI. Two rabbits were given parallel injections with the same batch of 3647 vaccine. Both rabbits produced strong antibodies against agglutinogens $a_4$ and $a_5$ while only rabbit K728 showed satisfactory titres against $b_1$ and $c_1$. Consequently both antisera could be used for the production of $a_4$

TABLE VI

**Antibody composition of two rabbit sera after immunisation with *S. aureus* strain 3647** (data from Haukenes, 1967)

| Antibodies | Serum K727 | Serum K728 |
|:---:|:---:|:---:|
| $a_4$ | + + + | + + + |
| $a_5$ | + + | + + |
| $b_1$ | + | + + + |
| $c_1$ | (+) | + + + |

factor serum while only serum K728 was suitable for the production of $c_1$ factor serum (see Table VII).

The result of immunisation depends, however, primarily on the selection of bacterial strains. The presence of a strong agglutinogen is no guarantee of a successful immunisation. The set of strains has been carefully selected and the strains will ordinarily produce satisfactory titres of the antibodies required. Nevertheless, at least two or three rabbits should be immunised simultaneously with the same vaccine to secure an optimal antiserum. A serum showing strong agglutination at a dilution of 1: 1000 to 1: 2000 with live 5 h bacteria of the homologous strain will usually give a satisfactory factor serum after absorption. After a successful immunisation, maximum amounts of serum should be drawn, since a booster injection may produce a different pattern of agglutinins. If the antibody response is not satisfactory after three series of injections further immunisation will usually not result in a good antiserum. The sera are preserved and stored as described on page 136. The titres usually keep fairly constant for years. If a serum is found to be less satisfactory after storage, a new one should immediately be produced.

### 3. *Factor sera*

The present typing set consists of 13 factor sera. Eleven are pure, i.e., containing antibodies against one single specific agglutinogen. Two sera are pools, each consisting of two types of antibodies. The pools are used because strains suitable for the production of pure $i_1$, $i_2$ and $k_2$ sera have not been found (Haukenes, 1964b). The absorption scheme is presented in Table VII.

Bacteria for absorption are grown on nutrient agar (Oxoid) plates 14 cm in diameter for 18 h at 37°C. Five to ten plates giving confluent growth are generally used for the absorption of 10 ml of diluted serum. The bacteria are suspended in saline with a Pasteur pipette and spun down. The supernatant is carefully pipetted off, 10 ml of a 1 : 10 dilution of antiserum

## TABLE VII

### Scheme for absorptions

| | |
|---|---|
| Factor serum $a_4$: | absorb serum 3647 w. strain F21, 1503, 17A, 3189. |
| Factor serum $a_5$: | absorb serum 17A w. strain 1503, 670. |
| Factor serum $b_1$: | absorb serum S365 w. strain C.I., autocl. S365. |
| Factor serum $c_1$: | absorb serum 3647 w. strain 2095. |
| Factor serum $h_1$: | absorb serum 17A w. strain 1503, 5687, C.I. |
| Factor serum $h_2$: | absorb serum 5687 w. strain 3647. |
| Factor serum $i_1i_2$ pool: | absorb serum F21 w. strain 1503, 3647. |
| Factor serum $k_1$: | absorb serum S365 w. strain F21, C.I. |
| Factor serum $k_1k_2$ pool: | absorb serum S365 w. strain F21. |
| Factor serum m: | absorb serum F21 w. strain 3647, W.46. |
| Factor serum n: | absorb serum 1503 w. strain 2095, C.I., 137. |
| Factor serum 263–1: | absorb serum 263 w. strain 3647, F21, 17A. |
| Factor serum 263–2: | absorb serum C.II w. strain 3647, 670. |

in saline is added and the bacteria are suspended in the serum by means of a Pasteur pipette. The tubes are placed in an incubator at 37°C for 2 h with intermittent shaking and then in the ice box overnight. The bacteria are spun down and the supernatant carefully pipetted off to avoid unclear serum due to bacteria or bacterial fragments. Sera requiring absorption with more than one strain are absorbed simultaneously with all the strains involved. This saves time and prevents loss and dilution of serum. Live bacteria are used for the absorptions, with one exception: Factor serum $b_1$ is produced by absorption with autoclaved S365 bacteria in order to remove antibodies against the heat-stable $k_1$ antigen, but not antibodies against the heat-labile $b_1$ antigen.

When the scheduled absorptions have been completed, the factor sera are checked preliminarily with the vaccine strain and the strain/strains used for absorption. Agglutination is performed on slides under the conditions indicated for each factor serum in the footnotes of Table VIII. If the absorption is incomplete it is repeated with the same amount of bacteria. Finally, when the factor sera have been exhausted of antibodies against the absorbing strains, they are checked by agglutination with all the 20 type strains. The agglutination patterns of the factor sera are given in Table VIII. A satisfactory factor serum should produce the strong agglutination reactions shown in the table and no reactions with strains by which negative reactions are indicated. Minor variations in weak reactions may occur and are unimportant. The titres of the factor sera vary from 1:500 to 1:10 with the different type strains (Haukenes and Oeding, 1960), usually being between 1:50 and 1:100.

Finally, the factor sera are heated at 60°C for 20 min, 2 drops of a 1% (w/v) solution of merthiolate are added to 5 ml portions of serum, and the

## TABLE VIII

### Agglutination patterns

| Type strain | Factor sera | | | | | | | | | | | | | Agglutination pattern |
|---|---|---|---|---|---|---|---|---|---|---|---|---|---|---|
| | $a_4$[2] | $a_5$[3] | $b_1$[1] | $c_1$[3] | $h_1$[1] | $h_2$[3] | $i_1i_2$[1] | $k_1$[3] | $k_1k_2$[3] | $m$[1] | $n$[1] | 263-1[1] | 263-2[1] | |
| 1503 | − | − | − | − | − | − | − | − | − | ++ | ++ | − | − | $m/n$ |
| 2253 | − | − | − | ++ | − | − | − | − | − | ++ | ++ | − | − | $c_1/m/n$ |
| 28 | − | − | − | − | − | − | − | − | − | − | ++ | − | − | $n$ |
| S365 | − | − | ++ | − | − | − | − | +++ | +++ | − | − | − | − | $b_1/k_1$ |
| 3647 | +++ | ++ | ++ | +++ | − | − | − | − | − | − | + | − | − | $a_4/a_5/b_1/c_1/n$ |
| F21 | − | − | +++ | +++ | − | − | +++ | − | − | +++ | − | − | − | $b_1/c_1/i_1i_2/m$ |
| 17A | − | +++ | − | − | + | + | − | − | − | − | − | − | + | $a_5/h_1/h_2/263\text{-}2$ |
| 3189 | − | − | − | − | − | − | − | − | − | − | − | − | − | NT |
| 2095 | ++ | ++ | ++ | − | − | − | − | − | − | − | − | − | − | $a_4/a_5/b_1$ |
| W.46 | − | − | − | − | − | − | ++ | − | − | − | − | − | − | $i_1i_2$ |
| C.I | − | − | − | − | − | ++ | − | − | ++ | ++ | − | ++ | + | $h_2/k_2/m/263\text{-}1/263\text{-}2$ |
| C.II | − | − | − | − | − | ++ | − | − | − | − | − | − | ++ | $h_2/263\text{-}2$ |
| C.III | +++ | ++ | ++ | +++ | − | − | − | − | − | − | + | − | − | $a_4/a_5/b_1/c_1/n$ |
| 263 | − | + | − | − | − | − | − | − | (+) | ++ | − | +++ | + | $a_5/k_2/m/263\text{-}1/263\text{-}2$ |
| 670 | − | − | − | − | + | +++ | − | − | − | − | − | − | − | $h_1/h_2$ |
| 830 | ± | +++ | − | − | − | − | − | − | − | − | − | − | − | $a_4/a_5$ |
| 1015 | − | − | − | ++ | − | − | − | − | − | − | − | − | − | $c_1$ |
| 5687 | − | − | − | − | − | ++ | − | − | − | − | − | − | + | $h_2/263\text{-}2$ |
| 6376 | − | − | − | +++ | − | − | − | − | − | − | − | − | − | $c_1$ |

*Culture for agglutination:* [1] 18 h nutrient agar 37°C. [2] 5 h nutrient agar 37°C. [3] 18 h mannitol-salt agar 37°C. Aggl. may be slightly granular.

sera are stored in the ice box in glass tubes with rubber stoppers. The titres usually remain unchanged for many months. However, if some time has elapsed since they were last used, the factor sera should be checked again. When possible the same batch of factor sera should be used for one typing project.

### 4. Agglutination

The purity of cultures of the coagulase-positive strains to be tested is checked on blood agar plates. If two types of colonies are seen in a culture, both are tested by agglutination. A pre-culture is made on a nutrient agar slant and after 18 to 24 h at 37°C a small inoculum is spread with a loop on:

1. A nutrient agar slant which is incubated for 18 h at 37°C;
2. A mannitol-salt agar slant which is incubated for 18 h at 37°C;
3. A nutrient agar slant which is incubated for 5 h at 37°C.

Eighteen hour nutrient agar cultures are used for agglutination in factor sera $b_1$, $h_1$, $i_1i_2$, $m$, $n$, 263–1 and 263–2, 5 h nutrient agar cultures for agglutination in factor serum $a_4$ and 18 h mannitol-salt agar cultures (Difco) for agglutination in factor sera $a_5$, $c_1$, $h_2$, $k_1$ and $k_1k_2$ (Table VIII). Five h nutrient agar and 18 h mannitol-salt agar cultures are used to avoid blocking of agglutination.

Agglutination is performed on slides using live bacteria. The slides are carefully wiped free of dust and film. One drop of factor serum is desposited with a Pasteur pipette on each half of a slide. The number of slides being examined simultaneously depends upon experience. Five to six slides (10–12 agglutinations) can easily be handled. The loop is filled with bacteria sufficient for all slides and a small amount is deposited beside each drop of factor serum. The bacteria are then gradually mixed with the serum by means of the loop until an even suspension is obtained. When all slides are ready they are agitated for some time by hand to accelerate agglutination. If necessary this is repeated while the reactions are read with the naked eye under a lamp against a dark background, and the results are recorded. When all slides have been examined the readings and the recorded results are checked. The agglutination reactions are graded: $+++$, $++$, $+$, $(+)$, $\pm$, $-$ and the result of serotyping is given as a pattern of the agglutinogens demonstrated, e.g. $c_1/m$ and $a_5/h_2/263$–1. Weak reactions $((+))$ are given in parentheses.

To ascertain reproducibility the experimental conditions should be kept as constant as possible. The culture media should be fresh and yield good growth. The temperature of the incubator should be checked at intervals since blocking of agglutination is more pronounced below 37°C (Oeding, 1957). The drops of factor sera as well as the amount of bacteria should be of about the same size each time to give suspensions of about the same density. Too thin and too thick suspensions should be avoided. The preparation of suspensions most convenient for reading is a matter of experience.

Each strain is checked for false clumping in a drop of saline. Self-agglutinating strains occur but do not constitute an important problem. Of the type strains, F21 has a tendency to self-agglutination. Agglutination is unreliable and should not be recorded if a strain shows some clumping in the saline control even if the clumping is weaker than in the factor sera. Self-agglutinating strains are retested in 5 h nutrient agar cultures, which provide more stable suspensions than older cultures (Haukenes, 1967). If the strain is still unstable, it is transferred to a new culture and retested. After one or more transfers a stable suspension is usually obtained.

Occasionally, false clumping may occur even if the saline control is completely stable (Haukenes, 1967). This is probably due to media and bacterial components dissolving in the factor sera during absorption. The experienced reader easily distinguishes false clumping from true agglutination. In false clumping the bacteria aggregate in the centre of the drop and have a tendency to sink below its surface. In true agglutination the bacteria become oriented peripherally in the drop and tend to float. Moreover, false clumping reaches a maximum very soon, whereas true agglutination gradually increases in strength on continuous agitation of the slide. Also, agglutination in many factor sera should direct suspicion towards false clumping, even if the saline control is stable. When this type of false clumping is suspected, saline should be replaced as a control by a normal rabbit serum, diluted 1 : 10, from which agglutinins have been removed by absorption. It should be mentioned that although mannitol-salt agar cultures give stable suspensions they are more granular than suspensions from nutrient agar cultures, but easily distinguishable from true agglutination in factor sera.

## C. Epidemiological value

For a serological method to be of epidemiological value several conditions have to be fulfilled. The typability must be good, a convenient number of types or patterns should be recognised, the results must be reproducible and the technique should not be too complicated. These conditions have been evaluated more or less carefully for the different *S. aureus* serotyping systems described since 1939 (Cowan). It must be admitted that efforts during the last 15 years have been aimed more at obtaining a basic knowledge of the agglutinogens, and at the technical details, than at epidemiological investigations. Consequently the revised typing system of Oeding has not been tested systematically with regard to the epidemiological conditions mentioned. However, an overall picture is obtained by studying the investigations performed through the years.

## 1. Typability

Non-typability may be due to blocking of the agglutinogens by other antigens or substances, to self-agglutination, or to the strain not being in possession of any of the agglutinogens represented in the typing set. Inagglutinability due to blocking has been eliminated as a problem by the introduction of mannitol-salt agar and other technical improvements. Non-typability due to self-agglutination is usually a minor problem which can be overcome by subcultures or autoclaving. Besides, it should be remembered that self-agglutination is in itself a characteristic which may contribute to the epidemiological evaluation.

The typability of human S. aureus strains is satisfactorily high with Oeding's method. In early investigations approximately 95% of the staphylococci were typable (Oeding, 1954; Grün, 1957). After the introduction of young (5 h) cultures for typing, the typability of 239 independent strains was 97% (Oeding and Williams, 1958) and in a recent investigation 98% (Fleurette and Modjadedy, 1976). Serotyping therefore identifies markers for most human strains of S. aureus including those non-typable by phages (Cohen, 1972; Live, 1972a). A large-scale investigation by Kretzschmar (1969) was more disappointing, showing a typability of only 82% of 700 human strains, which was lower than by phage typing. Although on the whole the new factor sera of Oeding – Haukenes were included, mannitol-salt agar cultures were not used systematically. This, in addition to the selection of the strains, probably explains the lower typability reported by Kretzschmar.

There seems to be a higher proportion of non-typable staphylococci from the air, skin and faeces than from pathological material (Grün, 1958b; Grün and Kühn, 1958). In an epidemiological investigation of an outbreak, the epidemic strain most probably has a well-defined serological pattern, and it is of minor importance if a few % of strains from carriers or the environment are non-typable because they are not of the epidemic type.

## 2. Number and occurrence of patterns

Because of the ubiquity of staphylococci it is important that a typing method is able to distinguish a reasonably large number of patterns or types. If the number is small the certainty with which an epidemic strain can be picked out in patients and carriers will suffer. On the other hand, it should be considered a disadvantage if a system registers a great number of patterns, unless they are well defined and not influenced by technical or other conditions (cf. phage typing).

The distribution of serological patterns in 239 independent human *S. aureus* strains was examined by Oeding and Williams (1958) using an early Oeding system. Approximately 35 different patterns were recorded but four patterns accounted for as much as 50% of the strains. Phage typing recognised a larger number of patterns, but this advantage had to be set against several disadvantages, such as the distinction between the patterns not always being clear. Similar findings have been reported by Grün (1959). The introduction of several new factor sera containing monovalent antibody permitted a greater number of patterns to be detected and a better differentiation of strains. Kretzschmar (1969), in an investigation of 700 human strains, was able to differentiate 156 serological patterns, some of which, however, showed only minor differences. Although few systematic investigations have been carried out, it may be concluded that the serological method of Oeding distinguishes a convenient number of patterns.

In his early work, Oeding (1952a, b; Oeding and Williams, 1958) attempted to collect strains with related antigenic patterns into types and groups. Kretzschmar (1969), in order to systematise serological typing, divided the agglutinogens into three orders according to their frequency and significance. Particularly Pillet (Pillet *et al.*, 1967), but also other authors (Sekiya, 1970), divide staphylococci into a fixed number of types according to their agglutination in a set of sera. An objection to this is that types should be based on knowledge of the antigens involved. Furthermore, the epidemiological value suffers, when the number of types is restricted.

## 3. *Reproducibility*

The reproducibility of serological typing can be evaluated either by repeated examinations of a number of colonies or subcultures of one strain, or by the examination of strains derived from the same source of infection. In such investigations, not only the technical procedure but also the antigenic stability as well as the possibility of double infection or contamination should be considered.

The reproducibility of the typing results connected with the technique depends primarily upon controlled and potent factor sera, rather than on standardised conditions for culture media, time, temperature and agglutination. In our experience there is a risk of non-specific reactions and reduced reproducibility if the sensitivity of the agglutination reaction is increased by reading with a hand lens or microscope (Grün, 1957; Kretzschmar, 1969), by keeping the slides in a moist chamber for a longer time (Grün, 1957), or by performing the agglutination in tubes (Stern and Elek, 1957). The standard technique is sufficiently sensitive to register all agglutinogens represented in the set, with the possible exception of antigens which are present in minimal amounts. Provided the technique is followed in every detail the reproducibility by repeated examinations is good. Although comparative typings of sets of strains have not been made in different laboratories there is reason to believe that the typing results of experienced workers would agree satisfactorily. It is, however, quite clear that direct comparison of results obtained using different methods cannot be made.

Serotyping is a rather reliable marker of a staphylococcal strain because of the stability of the agglutinogens. The type strains of Oeding's system have been kept in the laboratory and continuously checked for up to 25 years without changing their patterns of type agglutinogens. Pillet *et al.* (1950) and Pillet and Orta (1953) found Cowan's three type strains to be stable, and apart from two strains which lost their agglutinability but recovered it after passage in rabbits, their other strains showed no antigenic change. Grün (1957), Grün and Kühn (1958) and Kretzschmar and Kretzschmar (1962) also found that strains did not change their antigenic patterns over a longer observation period. Type agglutinogens of the human *S. aureus* ecotype are frequently found in strains of animal ecotypes, although their biochemical characteristics and phage sensitivities are different. This too is an indication of the stability of the staphylococcal agglutinogens (Oeding *et al.*, 1971). In contrast to these unanimous reports, Pereira (1961) is of the opinion that staphylococci regularly loose certain agglutinogens resulting in type variation.

Antibiograms and phage sensitivities are less stable than serological types. Antibiotic sensitivity may change with antibiotic usage and the emergence of resistant strains. Phage susceptibilities may change by interaction with wild phages by lysogenisation, by transduction, and by loss of a prophage. Stern and Elek (1957) reported that variants differing from the mother strains in pigment, haemolytic activity or sensitivity to antibiotics, retained their serologic types. Hofstad (1964b) observed that the antigenic pattern of *S. aureus* strain 263 remained unchanged after lysogenisation and repeated subcultures (Table IX). Cohen (1972) reported on experiments with *S. aureus* strain 502A. The laboratory introduction of a plasmid carrying penicillinase- and Erythromycin-resistance markers resulted in a change in phage type from (6)7 to 52/52A(80)/7, whereas the serotype remained $bc_1$. One example of documented serotypic changes has been reported with strain 502A which contained variants with serotype $a/b/c_1/m/n$ rather than $b/c_1$ (Cohen, 1972). The variants which had appeared earlier in the history of the culture and had been carried over in subculture, were discovered only after detailed study of cultures from a family into which the 502A strain had been introduced experimentally. In contrast,

TABLE IX

**Variant strains obtained by lysogenisation of strain 263**
(Hofstad, 1964b)

| Strain | Phage pattern | Antigenic pattern |
|---|---|---|
| 263 | 80/81/82/KS6 | $a_5(k_2)m/263-1/263-2$ |
| 263–80–1 | 80/81/82/KS6 | $a_5(k_2)m/263-1/263-2$ |
| 263–81–1 | 80 | $a_5(k_2)m/263-1/263-2$ |
| 263–81–2 | 52/52A/80/KS6 | $a_5(k_2)m/263-1/263-2$ |
| 263–KS6–3 | 52/52A/80/82 | $a_5(k_2)m/263-1/263-2$ |
| 263–KS6–4 | 52/52A/81/82 | $a_5(k_2)m/263-1/263-2$ |
| 263–KS6–6 | 52/52A/80/82 | $a_5(k_2)m/263-1/263-2$ |

phage typing variants of strain 502A are quite common (Cohen *et al.*, 1963a).

Several reports confirm the reproducibility of the Oeding system in epidemiological studies (Sompolinsky *et al.*, 1957; Oeding and Williams, 1958; Cohen *et al.*, 1963b; Cohen and Smith, 1965; Oeding *et al.*, 1973). Table X shows the results of serological typing in addition to phage typing of 11 strains of *S. aureus* isolated from different portions of a potato salad suspected to be the cause of an outbreak of staphylococcal food poisoning. The serological pattern is identical in all strains. Phage typing shows minor variations but clearly identifies all strains except No. 10 as being identical.

TABLE X

**Food strains of *S. aureus***
(data from Oeding and Sompolinsky, 1958)

| Strain No. | Antigenic pattern | Phage pattern |
|---|---|---|
| 1 | *a/b/c/h* | 6/7/47/75 + |
| 2 | *a/b/c/h* | 6/7/47/54/75 + |
| 3 | *a/b/c/h* | 6/7/47/75 + |
| 4 | *a/b/c/h* | 6/7/47/75 + |
| 5 | *a/b/c/h* | 6/7/47/54/75 + |
| 6 | *a/b/c/h* | 6/7/47/54/75 + |
| 7 | *a/b/c/h* | 6/7/47/75 + |
| 8 | *a/b/c/h* | 6/7/47/54/75 + |
| 9 | *a/b/c/h* | 6/7/47/54/75 + |
| 10 | *a/b/c/h* | NT |
| 11 | *a/b/c/h* | 6/7/47/54/75 + |

Factor sera: Oeding (1957)

TABLE XI

**Comparative typing of groups of strains thought to have a common source** (data from Oeding and Williams, 1957)

| Source of strains | No. of strains | Antigenic pattern |
|---|---|---|
| Food-poisoning outbreak | 8 | $a/b/c/h$ |
| Nose and septic lesion 1 person | 2 | $h$ |
| | 2 | $(a/h)$, $(h)$ |
| | 3 | Sp. aggl. |
| | 2 | $(a/b)e/h$, $(a)e/h$ |
| | 3 | $a/b/c/i$, $(a/b/c/i)$, $a/b/c(h)i$ |
| | 3 | $(a/b)e/h$, $(a)e/h$ |
| Hospital epidemic | 3 | $a/b/c/e/h$ |
| Food-poisoning outbreak | 3 | $i$ |
| Hospital epidemic | 3 | $(a/b/h)i$, $(a/b)i$ |
| Clothes of 1 individual | 4 | $h$ |
| Food-poisoning outbreak | 5 | $i, 1 = (b)i$ |

Factor sera: Oeding (1957)

The 41 strains presented in Table XI were selected with the special purpose of testing the reproducibility of the serological typing method of Oeding. The strains were selected on the basis of origin and identical reactions or minor variations in phage typing, indicating that the strains of each set probably had a common parent. When these strains were tested serologically the results were quite consistent, in that all the strains thought, from their origin and phage type, to have a common source, had identical or very similar serotypes.

Table XII shows an investigation of a group of cultures from a nursery

TABLE XII

**Cultures from nursery epidemic**
(data from Cohen, 1972)

| No. of strains | Antigenic pattern | Phage pattern |
|---|---|---|
| 13 | $a/b/c(h)k/m$ | 29/52/81(79,7) |
| 3 | $a/b/c/h/k/m$ | 52/81 |
| 1 | $a/b/c(h)k/m$ | 52/81(29, 79) |
| 1 | $a/b/c/k/m$ | 29/52/79/7/81 |
| 2 | $a/b/c/k/m$ | 29/52/81(79) |
| 2 | $a/b/c/k/m$ | 29/52/81/(79/7) |

Factor sera: Oeding (1957); Haukenes and Oeding (1960)

epidemic. Both phage typing and serotyping picked out the epidemic strain, although the typing results showed some variation in both methods. The consistency of serological typing was satisfactory, the only variation being due to the presence of a weak *h* antigen which was registered in the majority of the isolates but overlooked in some.

Tables XI and XII illustrate the problem which arises when weak serological reactions occur in epidemiological investigations. The problem is well known in phage typing in which the reading and interpretation of weak reactions has been standardised by agreement. It is evident that, by including the weak agglutination reactions, the number of typable strains is increased. The weak reactions also increase the number of patterns, thereby helping to distinguish strains in an investigation. However, the gain in sensitivity is to a certain degree accompanied by a loss in reproducibility. By recording weak reactions in parentheses the interpretation of slightly different patterns is facilitated.

## 4. *Comparison with phage typing*

The comparison of serological typing with phage typing is of practical interest with regard to correlation as well as to efficiency in epidemiological work, but also has theoretical significance concerning the receptor sites. Hobbs (1948) and Wahl and Fouace (1952) observed that certain phages showed a clear correspondence to Cowan's strains I, II and III. This led to the division of human *S. aureus* phages into three groups. A broad correlation between the major groups of serological typing and phage typing was subsequently confirmed by other authors (Oeding, 1953; Oeding and Vogelsang, 1954; Pillet *et al.*, 1954). Several later reports have focused on the correlation of the two systems, primarily with the object of combining them in epidemiological typing.

The comparison of single agglutinogens and phages is of special interest. Oeding and Williams (1958) observed that strains lysed by phage 187 possessed agglutinogen *k* either alone or in combination with other agglutinogens. This was confirmed by several authors (Cohen *et al.*, 1963b; Kretzschmar, 1969; Live, 1972b). Another striking correlation was detected by Hofstad (Hofstad and Oeding, 1962; Hofstad, 1964a) between agglutinogen 263–1 and the 80/81 phage complex. This focused the earlier reported broad correlation of Cowan strain I with phage group I on a single agglutionogen. The correlation has been corroborated in several reports (Cohen *et al.*, 1966; Kretzschmar, 1969; Galinski and Krynski, 1970). Less clear-cut correlations have been reported between agglutinogen *m* and phage group I (Haukenes, 1964b), between agglutinogen *a5* (Kretzschmar, 1969) as well as agglutinogen 263–2 (Galinski and Krynski, 1970) and phage group II, and between agglutinogen *a4* (Haukenes, 1963a, 1963c) respectively agglutinogens $n/a_4$ and phage group III (Kretzschmar, 1969).

The correlation between single agglutinogens and phages indicated that their receptor sites on the staphylococcal surface were shared or associated. A condition for this would be a complete correspondence between the serological demonstration of one particular agglutinogen and the sensitivity or adsorption, of one particular phage. Pillet *et al.* (1954) believed in such an association whereas Oeding and Williams (1958) concluded that the receptors were different. The correlation between agglutinogen $k$ and phage 187 has been the simplest one to study. Attempts to block the adsorption of either entity did not give conclusive results (Oeding, unpublished data). Recent investigations revealed that phage 187 strains contain a teichoic acid different from the glucosaminyl ribitol type usually present in strains of *S. aureus* (Karakawa and Kane, 1971; Oeding, 1974; Galinski and Lipinska, 1974). In accordance with the concept that the glucosaminyl ribitol teichoic acid is part of the phage receptor site of other *S. aureus* strains (Coyette and Ghuysen, 1968; Chatterjee, 1969), it is likely that the galactosaminyl ribitol teichoic acid serves this function in the 187 strains. It therefore seems unlikely that agglutinogen $k$ is the receptor substrate of phage 187 or that other agglutinogens are the substrates of other phages.

Phage typing is at present the method mostly used for the subdivision of *S. aureus*. This is not because it has been proved to be a better method than serotyping, but rather because it has been very well organized and standardized. Both methods can be used with success in epidemiological work. In one instance serotyping may provide the best information, in another instance phage typing, but as a rule the two methods are in good agreement. Illustrations of simultaneous typings are given in Tables X, XII, XIII and XIV. There are, however, some obvious advantages in serotyping, the most important being its high typability. The increasing number of phage non-typable strains during recent years has renewed interest in serotyping. Furthermore, the antigens are more stable than the phage markers. This is important for the reliability of typing and the permanency of the typing set. The greater similarity of the agglutinogens as compared to the phage sensitivity of *S. aureus* strains belonging to different ecotypes, makes serotyping with the human set more useful than the standard human set of phages for the typing of animal strains (Oeding *et al.*, 1970). In an epidemiological situation serotyping can be simplified. A limited number of factor sera can be used for screening and only the strains possessing the key agglutinogens of the epidemic strain need to be typed with the complete set.

The correlations demonstrated between phage typing and serotyping created an interest in combining the two systems in epidemiological investigations (Kretzschmar, 1969; Cohen, 1972). This approach involves several considerations. One method may be used for the first screening and the other for the detailed study of strains of particular epidemiological interest (Kretzschmar, 1969). For example, serology may be used for a finer differentiation of strains belonging to phage group I (Cohen, 1972). Few laboratories have the capacity for using both phage typing and serotyping and as a rule one method has to be chosen. It is a challenge to the advocators of serotyping to co-operate with the intention of standardising and simplifying serotyping to make it more widely accessible.

## 5. *Serotyping in outbreaks of staphylococcal infection*

Several epidemiological investigations using the early Oeding system showed that serotyping of *S. aureus* permits a reliable recognition of the epidemic strain and an evaluation of the background of hospital outbreaks (Grün, 1957, 1959; Kikuth and Grün, 1957; Sompolinsky *et al.*, 1957; Grün and Kühn, 1958; Lautermann and Grün, 1958; Oeding and Sompolinsky, 1958). Some authors have used strains, techniques and recording which differed considerably from the original method (Brodie *et al.*, 1955; Brodie, 1957; Löfkvist, 1957; Vischer, 1959; Sekiya, 1970). While no representative epidemiological investigation of human strains using the revised Oeding method in detail can be referred to, a few investigations have been performed with some of the new factor sera and technical procedures (Live and Nichols, 1965; Kretzschmar, 1969; Cohen, 1972). These investigations provide good evidence of the efficiency of the method.

In Table XIII we see results from a food poisoning episode. Here serotyping and phage typing were in complete agreement. Cultures from Cook II, the suspected food, and the victim had identical serotypes and phage patterns. The victim died and the affirmative results of two identification systems were of interest in legal questions concerning the incident (Cohen, 1972).

### TABLE XIII

**Cultures from food poisoning case**
(data from Cohen, 1972)

| Source | Antigenic pattern | Phage pattern |
| --- | --- | --- |
| Cook I | $c_1$ | 7(47, 53, 54, 75, 77) |
| Cook II | a/b/c | 6/47/53/83A(81) |
| Food | a/b/c | 6/47/53/83A(81) |
| Victim | a/b/c | 6/47/53/83A(81, 52A) |

Factor sera: Oeding (1957)

Table XIV presents data on cultures that were screened originally by phage typing in an outbreak of infections in a nursery due to a phage type 3A strain, and to smaller outbreaks of a similar nature in the same nursery at later times (Cohen, 1972). The cultures had been stored for a long time at room temperature but were nevertheless successfully serotyped. The $b/c_1/i/m$ and $c_1/i/m$ cultures were probably related. The two $c_1$ cultures in the first outbreak were isolated from the same nurse at different times. Some of the infections in the second and third outbreak were due to strains with the same serotype as those found in the first outbreak. These results again focus on the stability of the serotype and the relative transience of phage susceptibility. According to Cohen (1972), long range investigations like this one are almost impossible to carry out on the basis of phage typing alone.

TABLE XIV

**Outbreaks in a nursery**
(data from Cohen, 1972)

| No. of strains | Antigenic pattern | Phage pattern |
|---|---|---|
| | *First outbreak – July, 1961* | |
| 39 | $c_1/i/m$ | 3A |
| 20 | $b/c_1/i/m$ | 3A |
| 3 | $b/c_1/m$ | 3A |
| 2 | $c_1$ | 3A |
| | *Second outbreak – July, 1962* | |
| 5 | $b/c_1/i/m$ | 3A |
| 1 | $b/c_1/i/m$ | 3A/71 |
| 1 | $b/c_1/i/m$ | 3A/3B/3C |
| 2 | $c_1$ | 3A |
| 1 | $a/b/c/h/k/m$ | 3A/3B/3C/55/71 |
| 1 | $k/m$ | 3A/3B/3C/55/71 |
| 1 | $a/b/c/h/k$ | 55 |
| 1 | $h/k$ | 29/52 |
| | *Third outbreak – February, 1963* | |
| 2 | $b/c_1/i/m$ | NT |
| 1 | $c_1/i/m$ | NT |
| 1 | $b/c_1$ | NT |
| 1 | $b/h/k/m$ | 3A/3B/3C/55/71 |
| 1 | $b/c/k$ | NT |

Factor sera: Oeding (1957); Haukenes and Oeding (1960)

Tables X, XI and XII illustrate outbreaks of staphylococcal infections in which serotyping has been used with success.

## D. Animal strains

In epidemiological work on animal staphylococci, the first problem is to distinguish strains belonging to different biotypes (Meyer, 1967; Hájek and Maršálek, 1971). This is necessary since cross-colonisation occurs quite frequently. A subdivision of the strains of one biotype is also desirable. Both phage typing, using the basic sets of phages for typing human and bovine staphylococci, and serotyping, may supplement the biochemical and cultural methods in determining the biotype and may also successfully subdivide the strains of one biotype. Serotyping at present provides the best information (Oeding et al., 1973, 1974), but it must be remembered that, with the exception of phage typing of bovine strains, both methods are provisional for typing animal staphylococci. Serological determination of the type of teichoic acid and protein A supplement the characterisation (see p. 140).

The sera used for serotyping of animal S. aureus strains are produced against human staphylococci. Early papers on the typing of animal strains gave rather poor results (Pillet et al., 1950; Mercier et al., 1950; Grün, 1958a), indicating that animal and human staphylococci possess different agglutinogens. However, the increasing knowledge of the human S. aureus agglutinogens and the perfecting of the typing set revealed that some specific agglutinogens were shared and that, after all, the human typing set was quite useful in typing animal staphylococci. In bovine strains from infections and carriers a typability of 65–99% was reported (Malik and Singh, 1960; Slanetz and Bartley, 1962; Marandon and Oeding, 1967; Oeding et al., 1971). Canine staphylococci have been reported to be typable in 0% (Shimizu, 1968), 45% (Pillet et al., 1968) and up to 85% (Live and Nichols, 1965; Hahn and Blobel, 1968; Oeding et al., 1970). High typability has also been reported in strains from sheep (Pillet et al., 1962; Plommet and Wilson, 1969; Oeding et al., 1976), swine (Oeding et al., 1972) and hare (Oeding et al., 1973), whereas the typing of strains from horses (Ochi and Shimizu, 1960; Shimizu, 1968; Yoshimura, 1970; Oeding et al., 1974), pigeons (Oeding et al., 1970) and mink (Oeding et al., 1973) has given poorer results.

While serological typing of human strains regularly reveals a number of agglutinogens, only one or two agglutinogens are as a rule demonstrated in animal strains using the human typing set (Grün, 1958a, Pulverer, 1965; Marandon and Oeding, 1967; Oeding et al., 1976). Very often, however, the single agglutinogen demonstrated in animal strains is very strong. According to several reports (Slanetz and Bartley, 1962; Marandon and

Oeding, 1967; Oeding et al., 1971; Oeding et al., 1972; Oeding et al., 1973; Oeding et al., 1976), the $c_1$ and $h_2$ agglutinogens are the human agglutinogens most frequently present in animal strains. Next to these, the $a_5$ agglutinogen has been demonstrated (Oeding et al., 1970), and especially in some canine strains, the $k$ agglutinogen (Oeding et al., 1970; Live, 1972b).

This shows that a limited number of specific agglutinogens are shared by human and animal S. aureus strains. There is, at present, no evidence for the correlation of a particular agglutinogen to an animal species. The rare occurrence of human agglutinogens indicates that animal staphylococci possess their own specific agglutinogens. Systematic research has not been done but improved typability using absorbed sera against canine (Shimizu, 1968; Pillet et al., 1968), horse (Ochi and Shimizu, 1960; Shimizu, 1968; Yoshimura, 1970) and bovine strains (Kretzschmar, 1976) has been reported. Only by systematic antigen analyses of strains belonging to all biotypes, can serotyping systems giving maximum yield be expected.

An example of successful serotyping of strains isolated from the nares of healthy dogs is shown in Table XV. No strain belonging to the canine biotype was typable with the basic sets of phages for typing human and bovine strains, whereas 85% were easily typed serologically. Fourteen antigenic patterns were recorded, 75% of the strains belonging to three

TABLE XV

**Serological typing of 75 dog carrier strains**
(Oeding et al., 1970)

|  | Antigenic pattern | No. of strains |
|---|---|---|
|  | $a_5/c_1/k_1$ | 19 |
|  | $a_5/k_1$ | 13 |
| 63 strains of | $a_5$ | 9 |
| canine biotype | $a_5$/other antigens | 8 |
| and 2 | $k_1$ | 1 |
| intermediate | $h_1/m/263$–1 | 4 |
| strains | $c_1/k_1$ | 1 |
|  | NT | 10 |
| 10 strains of | $a_5$/other antigens | 2 |
| human biotype | Different patterns without $a_5$ antigen | 8 |
| Total |  | 75 |

NT = non-typable

patterns. Of ten strains having the characteristics of the human biotype, eight were typable by phage and all were typable serologically, showing patters different from those of the canine biotype. $a_5$ was the strong and dominating agglutinogen of the canine biotype strains, the human biotype strains containing several agglutinogens. In this investigation, therefore, both methods picked out the strains transmitted from human beings but only serotyping permitted a differentiation of the canine biotype strains.

In an investigation of 21 *S. aureus* strains isolated from the nares of hares, both phage typing and serotyping were quite successful (Table XVI). Both methods agreed in that all strains of the hare biotype were of the same type. In these strains, agglutinogen $h_2$ occurred alone, giving strong agglutination. The four E biotype strains, probably transmitted from pointers, were non-typable by phage, whereas serotyping indicated three different sources. Two strains contained unusually many agglutinogens.

TABLE XVI

**Serological and phage typing of 21 *S. aureus* strains isolated from hares**
(data from Oeding *et al.*, 1973)

| Biotype | Phage pattern | Serological pattern | No. of strains |
|---------|---------------|---------------------|----------------|
| D | 3A/3C/55/71/116 | $h_2$ | 17 |
| E | NT | $a_5$ | 2 |
|  |  | $a_5/e/h_1/h_2/k_1k_2$ | 1 |
|  |  | $a_5/e/h_1/k_1k_2/n/263$–1 | 1 |

## E. Perspectives

If serological typing of *S. aureus* is to come into general use, the simplification of the method must be an aim of future work. The procedures should be simplified and standardised in so far as this can be done without reducing the sensitivity and specificity of the method. The production of antisera and factor sera is time-consuming and the sera may not, in spite of control measures, be absolutely comparable from laboratory to laboratory. Centralisation of typing sera production under expert control would be an important step towards standardisation and simplification. Furthermore, a reference laboratory for serotyping staphylococci would be very valuable. The reference laboratory would have to organise compared typing of strains in participating laboratories, check and deliver the type strains agreed upon, give advice and organise discussions.

Although the method of Oeding is today used in nearly all laboratories performing serotyping of *S. aureus*, several minor modifications complicate the comparison of results. Pillet *et al.* (1967) proposed procedures to make serotyping more practical and made suggestions for an amalgamation of Pillet's system and that of Oeding. An interesting contribution towards standardisation was recently given by Fleurette and Modjadedy (1976). The modifications proposed by them should be examined, and serious attempts should now be made to agree internationally upon every detail of one single typing method.

Furthermore, the need for serotyping animal staphylococci should encourage systematic analyses of their agglutinogens. The poor knowledge of the chemical nature of the type agglutinogens is an obvious disadvantage compared to other typing systems such as those developed for pneumococci, streptococci and enteric bacteria. Efforts should therefore be made to isolate and characterise the agglutinogens.

## VI. SUBDIVISION OF *S. EPIDERMIDIS* AND *S. SAPROPHYTICUS* BY AGGLUTINATION

The type of teichoic acid and the lack of protein A are valuable criteria for the species/biotype differentiation of coagulase-negative staphylococci and their separation from *S. aureus*. Serological methods for this have been described on page 141. Although the agglutinogens of coagulase-negative staphylococci are known to be different from those of *S. aureus*, no serological method exists for the typing of the former. The present state of knowledge will, however, be shortly reviewed.

Cross-agglutination between *S. epidermidis* and *S. aureus* occurs regularly (Pillet and Orta, 1954; Morse, 1962; Aasen and Oeding, 1971). Some antisera have rather high titres of cross-reacting antibodies, whereas in others the titres are relatively low. There are at least two, and probably several, specificities among the shared agglutinogens (Aasen and Oeding, 1971), the nature of which are unknown.

It should be noticed that rabbit pre-immune sera contain antibodies against *S. aureus* as well as *S. epidermidis* agglutinogens, although usually in low titres (Aasen and Oeding, 1971). If factor sera for typing *S. aureus* are produced without proper measures for removing pre-immune antibodies, shared antibodies may elicit agglutination of *S. epidermidis* strains and give the false impression that they contain specific *S. aureus* agglutinogens (Grün, 1965; Reiss *et al.*, 1969). It has, however, been definitely shown that the type agglutinogens of *S. aureus* are specific for this species (Haukenes, 1967; Aasen and Oeding, 1971).

Pillet and Orta (1970) attempted serotyping coagulase-negative staphylococci with absorbed antisera against two serologically distinct strains as well as with sera for their 13 *S. aureus* types. By this procedure they were able to type about 50% of 194 coagulase-negative strains. Similar results were obtained by Brun (1971). These investigations demonstrated the existence of specific *S. epidermidis* agglutinogens, but the antisera also contained antibodies against agglutinogens shared with *S. aureus*, and the presence of pre-immune antibodies was not considered.

A preliminary analysis of the specific agglutinogens of *S. epidermidis* was reported by Aasen and Oeding (1971). After removal of shared antibodies by absorption of the *S. epidermidis* antisera with bacteria of selected *S. aureus* strains, the remaining antibodies were identified by further absorptions with *S. epidermidis* strains. Several specific *S. epidermidis* agglutinogens were recorded, some being present in only one strain, whereas others were shared by several strains. Similar results were obtained by Hasselgren and Oeding (1972) in an examination of the urinary strains of *S. saprophyticus* ("Micrococcus" biotype 3). The results indicated that the number and quantities of the specific agglutinogens are lower in *S. epidermidis* than in *S. aureus*.

Using this approach, which is identical with that of Oeding for typing *S. aureus* strains, there is hardly any doubt that a satisfactory serotyping system can be obtained also for *S. epidermidis*. Monospecific typing sera against specific *S. epidermidis* agglutinogens should be used for the serological subdivision of clearly defined species or biotypes.

## VII. PHAGE TYPING

For the subdivision of *S. aureus* strains of human and bovine origin, phage typing is the prevailing method. Phage typing of *S. aureus* was reviewed as recently as 1972 by Parker in Vol. 7B of this Series. Only the more recent developments will, therefore, be discussed here.

### A. International basic set for typing *S. aureus* strains of human origin

At the 6th meeting of the International Subcommittee on Phage typing of Staphylococci (1975), several recommendations with consequences for the composition and use of the international basic set of phages for typing *S. aureus* strains of human origin were made. The Subcommittee agreed that the use of an international standard typing medium, though desirable, was not possible at present due to shortage of agar. It was, however, urged that, as far as possible, the same medium be used in individual countries.

The percentage of strains of *S. aureus* that can be typed with the basic set of phages has fallen in recent years. It is therefore essential that new phages be introduced if the situation is to be improved. Several phages have been examined as possible candidates and their value assessed. It is advisable that the basic set of phages should be kept at 23 to 24. Because the existing set consists of 22 phages, the addition of new phages had to be followed by the removal of old ones. The Subcommittee recommended that three phages, numbers 94, 95 and 96, be added to the basic set and that phages 42D and 187 be removed, but that these two phages be maintained in national laboratories as extra phages to be used when required. The new basic set thus consists of 23 phages:

Group I:   29, 52, 52A, 79, 80
Group II:  3A, 3C, 55, 71
Group III: 6, 42E, 47, 53, 54, 75, 77, 83A, 84, 85
Not allocated: 81, 94, 95, 96

It has been the rule since 1966 that new phages should be used only at RTD and not at RTD × 1000 or RTD × 100. This practice led to a decrease in typability and also made it difficult to recognise epidemiologically related strains. The Subcommittee, therefore, decided that in future all phages should be used at these concentrations.

A number of phages that are not in the basic set are useful in identifying non-typable strains in particular countries, and laboratories have been encouraged to use such locally useful phages. It was the rule, however, that reactions obtained with them could only be reported when the strain was non-typable with the basic set of phages at both RTD and RTD × 100. This being uneconomical of both time and material it was decided that in future reactions with additional phages may be accepted without the need to re-type strains with basic-set phages at RTD × 100.

## B. Animal *S. aureus* strains

The basic set of phages for typing human staphylococci has not been found generally applicable to animal *S. aureus* strains. Two reasons are given for this (Parker, 1972): Most of the animal strains are not lysed by the human phages at RTD, or those that are typable give wide patterns that are unstable and difficult to interpret. Non-typability has particularly been observed with staphylococci from dogs, pigeons (Oeding *et al.*, 1970), horses (Oeding *et al.*, 1974), mink (Oeding *et al.*, 1973) and sheep (Oeding *et al.*, 1976). It is clear that independent phage typing systems would be necessary for the study of staphylococcal disease in several animal species. Investigations of the development of such phages have so far been concerned mainly with cultures from bovine and canine (Coles, 1963) species. An international basic set of phages for typing *S. aureus* strains of bovine origin has been accepted by the Subcommittee (1971). The bovine basic set includes 9 members of the present human basic set; the remaining phages include one former "additional" phage from the human set and 6 phages of bovine origin. The bovine set has proved valuable not only for the typing of strains from cattle, but also as a supplement to the human basic set for typing strains from other animal species. Of 75 sheep strains of biotype C, 95% were sensitive to bovine phage 78 (Oeding *et al.*, 1976), sensitivity to this phage apparently being a characteristic shared by the majority of sheep staphylococci (Davidson, 1961).

## C. *S. epidermidis*

A phage typing system for *S. epidermidis* has recently been described by Verhoef *et al.* (1972) and at present several groups are actively engaged in this work. The need for typing methods in the epidemiological evaluation of serious systemic infections caused by *S. epidermidis* strains has been pointed out above (see p. 128).

## VIII. BACTERIOCIN TYPING

Fredericq (1946) introduced the name staphylococcins for bacteriocins produced by staphylococci. He observed that staphylococci produce a number of substances that inhibit the growth of other staphylococci and some other Gram-positive bacteria, but not Gram-negative bacteria. Less work has been done on bacteriocins from Gram-positive bacteria than on antibacterial substances produced by members of the family *Entero-bacteriaceae*. Several staphylococcins have, however, been described which have different antibacterial spectra and have been given different names (epidermidin, micrococcin, aureocin). The data given are often conflicting and it is difficult to evaluate the relationship between these substances.

Staphylococcins have most commonly been found in phage group II strains, particularly in strains lysed by phage 71 alone (Parker *et al.*, 1955; Fink and Ortel, 1969; Dajani and Wannamaker, 1973). Dajani and Wannamaker (1973) reported that this bactericidal substance was produced by 34% of 75 strains lysed by either phage 71 alone or other phages in group II in addition to phage 71. Phage types in group II other than type 71, phage types in other groups and strains of *S. epidermidis* did not produce the substance. The agent was first recognised by Parker *et al.* (1955), who demonstrated its powerful action on *Corynebacterium diphtheriae in vitro*. It was later shown to inhibit all strains of *S. aureus* that did not produce it, as well as a wide range of other Gram-positive bacteria, but not *S. epidermidis* (Dajani and Wannamaker, 1969, 1973). Among sensitive organisms, definite quantitative variations exist in the susceptibility to the substance.

The isolation and purification of this substance as well as its mode of action have been studied by Dajani *et al.* (1970) and Dajani and Wannamaker (1973). The substance is an extracellular product, heat stable, of protein or polypeptide nature and kills but does not lyse bacteria. According to Richmond (1972), it is likely that the genes responsible for bacteriocin production in staphylococci are extrachromosomal and in many ways analogous to the colicinogenic factors detected in the enteric bacteria. It is suggested that bacteriocin sythesis in staphylococci is specified by a plasmid.

Another antibacterial substance called staphylococcin A was isolated by Lachowicz (1965) from a non-pathogenic staphylococcus. Of 1239 strains of staphylococci, 10·6% showed different degrees of antagonism, indicating that several antibacterial substances were involved. Whereas 42% of the strains were susceptible to phage 71, the remaining strains belonged to other phage types or were non-typable. Both coagulase-positive and coagulase-negative strains were active. Several other authors (Fink and Ortel, 1969; Gagliano and Hinsdill, 1970; Moore, 1970; Hale and Hinsdill, 1975) have reported on antibacterial substances, the relationship of which to the agents described above is difficult to evaluate.

As early as in 1946, Fredericq suggested that *Enterobacteriaceae* species might be subdivided by means of colicins analogous to phage typing. This was attempted but without success. However, a subdivision of *Shigella sonnei*, *Escherichia coli* and *Pseudomonas aeruginosa* can be achieved by testing the bacteriocinogenic activity of unknown strains against standard indicator strains, different strains producing different inhibition patterns. This method has been used to some extent.

In contrast to the highly specific and limited spectra of bacteriocins of Gram-negative bacteria, those of Gram-positive bacteria have a wide range of activity. The implications of this for a staphylococcin typing system is not quite clear. Hale and Hinsdill (1975) found it easier in other genera than within the staphylococci, to detect strains susceptible to their staphylococcin 462. At present no such system is used for typing staphylococci or micrococci. The frequent production of staphylococcin by phage 71 strains and the infrequent production by other *S. aureus* strains, might suggest bacteriocin typing as a supplement to phage typing and serotyping in the subdivision of phage group II. It should be mentioned that Moore (1970) suggested that the "aureocin" demonstrated by her might be used for the differentiation between pathogenic and non-pathogenic staphylococci. The applicability of bacteriocin typing on staphylococci must await more knowledge of the production and specificities of the staphylococcins.

Lysostaphin, described by Schindler and Schuhardt (1964), is different from the bacteriocins. Lysis is caused by the specific action of an endopeptidase on the staphylococcal peptidoglycan. Sensitivity to lysostaphin is used as a criterion for the differentiation of staphylococci and micrococci (Lachica et al., 1971; Schleifer and Kloos, 1975b).

# APPENDIX

## A. Genus *Micrococcus*

In the 8th edition of Bergey's Manual (Baird-Parker, 1974a) the number of *Micrococcus* species has been reduced to three from 16 in the 7th edition. These species are: *M. luteus*, *M. roseus* and *M. varians*. New species have later been described by Kloos *et al.* (1974). Baird-Parker subdivided genus *Micrococcus* into 8 biotypes (1963) but his scheme is not valid after the transfer of biotype 3 and some additional strains to genus *Staphylococcus* (*S. saprophyticus*). The $G+C$ content of the DNA of micrococci ranges from 66-75 moles % (Baird-Parker, 1974a). Micrococci do not contain teichoic acids (Schleifer and Kandler, 1972; Baird-Parker, 1974b) and are characterised by peptidoglycan types other than those found in staphylococci (Schleifer and Kandler, 1972). The inability to grow and produce acid anaerobically from glucose is the main laboratory criterion for the differentiation of micrococci from staphylococci. The unreliability of this test may be the reason why some authors consider the micrococcus a potential pathogen.

Very little is known of the antigens of micrococci. Although micrococci are said to have no teichoic acids it is interesting that approximately 50% of *Micrococcus* strains isolated from urinary samples were found by Digranes and Oeding (1975) to produce the poly C line by agar precipitation. This suggests that these micrococci contain a teichoic acid identical or similar to the N-acetyl $\beta$-glucosaminyl glycerol teichoic acid of poly C (Johnsen *et al.*, 1975a, b), or some other cross-reacting polysaccharide material. In an earlier examination of 20 culture collection strains classified as *M. luteus*, *M. roseus*, *M. conglomeratus* and *M. varians*, four strains cross-reacted with poly $A_{\beta}$ in agar precipitation (Oeding, 1967). Cross-reaction between poly $A_{\beta}$ and poly C later having been demonstrated (Johnsen *et al.*, 1975a, b), the four strains probably contained poly C. Three strains cross-reacted with *Micrococcus* strain I3 which contains an $\alpha$-glucosaminyl glycerol teichoic acid (Oeding *et al.*, 1967).

The same 20 strains of micrococci were examined for their content of agglutinogens (Oeding, 1967). Cross-absorption experiments showed that they possessed shared as well as more or less specific agglutinogens. Some strains had similar antigenic patterns with a certain correlation to the species. The antigenic composition of the micrococcal strains revealed a clear relationship to *S. epidermidis* and even to *S. aureus*. Staphylococci seem to have a common basic structure of agglutinogens, which to some extent is shared also by micrococci. In addition each staphylococcal species and micrococci contain specific agglutinogens. Further work on the specific agglutinogens of micrococci should be performed with typical strains classified by modern criteria in approved species. This applies also to experiments on phage typing (Peters *et al.*, 1976).

## B. Genus *Planococcus*

Most authors have previously placed the flagellated, Gram-positive cocci of marine origin in genus *Micrococcus*. In the 8th edition of Bergey's Manual (Kocur, 1974), they are recognised as Genus III, *Planococcus* of the family *Micrococcaceae*. This was done primarily because they differed clearly from staphylococci and micrococci in their % guanine plus cytosine (GC) content (Boháček and Kocur, 1968). The separation from micrococci was supported by Schleifer and Kandler

(1970), who found that the strains studied were uniform with respect to the type of murein present in their cell walls, and that this murein was different from the mureins found in the cell walls of members of the genera *Micrococcus* and *Staphylococcus*.

An antigenic study was performed by Oeding (1971) on the same strains used by the authors cited above. The precipitinogens and agglutinogens demonstrated revealed no relationship to staphylococcal or micrococcal antigens, thus substantiating the conclusion that these flagellated cocci do not belong to genus *Micrococcus*. Antigenically the strains were rather heterogeneous, but in accordance with their % GC content two groups were recorded.

## REFERENCES

Aasen, J. and Oeding, P. (1971). *Acta path. microbiol. scand. Sect. B*, **79**, 827–834.

Andersen, E. K. (1943). *Acta path. microbiol. scand.*, **20**, 242–256.

Archibald, A. R. (1972). *In* "The Staphylococci" (J. O. Cohen, Ed.), pp. 75–109. Wiley–Interscience, New York, London, Sydney and Toronto.

Armstrong, J. J., Baddiley, J., Buchanan, J. G., Carss, B. and Greenberg, G. R. (1958). *J. chem. Soc.*, **1958**, 4344–4354.

Baddiley, J., Buchanan, J. G., Hardy, F. E., Martin, R. O., RajBhandary, U. L. and Sanderson, A. R. (1961). *Biochim. biophys. Acta*, **52**, 406–407.

Baird-Parker, A. C. (1963). *J. gen. Microbiol.*, **30**, 409–427.

Baird-Parker, A. C. (1974a). *In* "Bergey's Manual of Determinative Bacteriology" (R. E. Buchanan and N. E. Gibbons, Eds), 8th edn., pp. 478–489. The Williams and Wilkins Company, Baltimore.

Baird-Parker, A. C. (1974b). *Ann. N.Y. Acad. Sci.*, **236**, 7–13.

Boháček, J. and Kocur, M. (1968). *J. appl. Bact.*, **31**, 215–219.

Brodie, J. (1957). *J. clin. Path.*, **10**, 215–218.

Brodie, J., Jamieson, W. and Sommerville, T. (1955). *Lancet*, ii, 223–225.

Brun, Y. (1971). "Staphylococcus epidermis". Thesis. Editions de l'Association Corporative des Etudiants en Médicine de Lyon.

Chatterjee, A. N. (1969). *J. Bact.*, **98**, 519–527.

Christie, R. and Keogh, E. V. (1940). *J. Path. Bact.*, **51**, 189–197.

Cohen, J. O. (1972). *In* "The Staphylococci" (J. O. Cohen, Ed.), pp. 419–430. Wiley-Interscience, New York, London, Sydney and Toronto.

Cohen, J. O. and Smith, P. B. (1965). *Hlth Lab. Sci.*, **2**, 197–202.

Cohen, J. O., Smith, P. B., Shotts, E. B., Boris, M. and Updyke, E. L. (1963a). *Am. J. Dis. Child.*, **105**, 689–691.

Cohen, J. O., Smith, P. B. and West, B. (1963b). *Bact. Proc.*, p. 85, Am. Soc. Microbiol.

Cohen, J. O., Smith, P. B., Erwin, M. L. and Fix, R. M. (1966). *J. Bact.*, **91**, 2108–2109.

Coles, E. H. (1963). *Am. J. vet. Res.*, **24**, 803–807.

Cowan, S. T. (1939). *J. Path. Bact.*, **48**, 169–173.

Coyette, J. and Ghuysen, J. M. (1968). *Biochemistry*, **7**, 2385–2389.

Dajani, A. S. and Wannamaker, L. W. (1969). *J. Bact.*, **97**, 985–991.

Dajani, A. S. and Wannamaker, L. W. (1973). *In* "Staphylococci and Staphylococcal Infections". Proc. 2nd Int. Symp., Warsaw, 1971 (J. Jeljaszewicz, Ed.), pp. 413–421. Polish Med. Publ., Warsaw.

Dakani, A. S., Gray, E. D. and Wannamaker, L. W. (1970). *J. exp. Med.*, **131**, 1004–1015.

Davidson, I. (1961). *Res. vet. Sci.*, **2**, 396–407.

Davison, A. L. and Baddiley, J. (1963). *J. gen. Microbiol.*, **32**, 271–276.

Davison, A. L. and Baddiley, J. (1964). *Nature, Lond.*, **202**, 874.

Davison, A. L., Baddiley, J., Hofstad, T., Losnegard, N. and Oeding, P. (1964). *Nature, Lond.*, **202**, 872–874.

Devriese, L. and Oeding, P. (1975). *J. appl. Bact.*, **39**, 197–207.

Devriese, L. and Oeding, P. (1976). *Res. vet. Sci.*, **21**, 284–291.

Digranes, A. and Oeding, P. (1975). *Acta path. microbiol. scand. Sect. B*, **83**, 373–381.

Ekstedt, R. D. (1974). *Ann. N. Y. Acad. Sci.*, **236**, 203–220.

Elliot, S. D., Gillespie, E. H. and Holland, E. (1941). *Lancet*, i, 169–171.

Endresen, C., Grov, A. and Oeding, P. (1974). *Acta path. microbiol. scand. Sect. B*, **82**, 382–386.

Endresen, C. and Grov, A. (1976). *Acta path. microbiol. scand. Sect. B*, **84**, 305–308.

Evans, J. B. and Kloos, W. E. (1972). *Appl. Microbiol.*, **23**, 326–331.

Fink, H. and Ortel, S. (1969). *Zentbl. Bakt. ParasitKde, Abt. I Orig.*, **211**, 39–47.

Fleurette, J. and Modjadedy, A. (1976). *In* "Staphylococci and Staphylococcal Diseases". Proc. 3rd Int. Symp., Warsaw, 1975. (F. Feljaszewics, Ed.), pp. 71–80. Fisher, Stüttgart.

Forsgren, A. and Nordström, K. (1974). *Ann. N. Y. Acad. Sci.*, **236**, 252–266.

Forsgren, A. and Sjöquist, J. (1966). *J. Immun.*, **97**, 822–827.

Fredericq, P. (1946). *C. r. Séanc. Soc. Biol.*, **140**, 1167–1170.

Gagliano, V. J. and Hinsdill, R. D. (1970). *J. Bact.*, **104**, 117–125.

Galinski, J. and Krynski, S. (1970). *Pathologia Microbiol.*, **35**, 393–400.

Galinski, J. and Lipinska, E. (1974). *J. Hyg. Epidem. Microbiol. Immun.*, **18**, 219–225.

Gillespie, E. H., Devenish, E. A. and Cowan, S. T. (1939). *Lancet*, ii, 870–873.

Grov, A. (1969a). *Acta path. microbiol. scand.*, **76**, 621–628.

Grov, A. (1969b). *Acta path. microbiol. scand.*, **76**, 629–636.

Grov, A. and Oeding, P. (1971). *Acta path. microbiol. scand. Sect. B*, **79**, 539–544.

Grov, A. and Rude, S. (1967). *Acta path. microbiol. scand.*, **71**, 417–421.

Grov, A., Myklestad, B. and Oeding, P. (1964). *Acta path. microbiol. scand.*, **61**, 588–596.

Grov, A., Myklestad, B. and Oeding, P. (1966). *Acta path. microbiol. scand.*, **68**, 149–156.

Grün, L. (1957). *Z. Hyg. InfektKrankh.*, **144**, 238–247.

Grün, L. (1958a). *Z. Hyg. InfektKrankh.*, **145**, 259–262.

Grün, L. (1958b). *Arch. Hyg. Bakt.*, **142**, 3–7.

Grün, L. (1959). *Z. Hyg. InfektKrankh.*, **146**, 129–141.

Grün, L. (1961). *Zentbl. Bakt. ParasitKde, Abt. I Orig.*, **181**, 210–215.

Grün, L. (1965). *Zentbl. Bakt. ParasitKde, Abt. I Orig.*, **195**, 321–325.

Grün, L. and Kühn, H. (1958). *Z. Hyg. InfektKrankh.*, **144**, 535–548.

Hahn, G. and Blobel, H. (1968). *Zentbl. VetMed.*, **15**, 979–983.

Hájek, V. and Maršálek, E. (1969). *Zentbl. Bakt. ParasitKde, Abt. I Orig.*, **212**, 60–73.

Hájek, V. and Maršálek, E. (1971). *Zentbl. Bakt. ParasitKde, Abt. I Orig. A*, **217**, 176–182.

Hale, E. M. and Hinsdill, R. D. (1975). *Antimicrobial Agents and Chemotherapy*, **7**, 74–81.

Hasselgren, I. L. and Oeding, P. (1972). *Acta path. microbiol. scand. Sect. B*, **80**, 257–264.

Haukenes, G. (1962a). *Acta path. microbiol. scand.*, **55**, 117–126.

Haukenes, G. (1962b). *Acta path. microbiol. scand.*, **55**, 450–462.

Haukenes, G. (1962c). *Acta path. microbiol. scand.*, **55**, 463–474.

Haukenes, G. (1963a). *Acta path. microbiol. scand.*, **59**, 205–212.

Haukenes, G. (1963b). *Acta path. microbiol. scand.*, **59**, 213–219.

Haukenes, G. (1963c). *Acta path. microbiol. scand.*, **59**, 220–228.

Haukenes, G. (1964a). *Acta path. microbiol. scand.*, **60**, 285–294.

Haukenes, G. (1964b). *Acta path. microbiol. scand.*, **61**, 283–290.

Haukenes, G. (1964c). *Acta path. microbiol. scand.*, **61**, 415–426.

Haukenes, G. (1967). *Acta path. microbiol. scand.*, **70**, 590–600.

Haukenes, G. and Oeding, P. (1960). *Acta path. microbiol. scand.*, **49**, 237–248.

Haukenes, G., Ellwood, D. C., Baddiley, J. and Oeding, P. (1961a). *Biochim. biophys. Acta*, **53**, 425–426.

Haukenes, G., Losnegard, N. and Oeding, P. (1961b). *Acta path. microbiol. scand.*, **53**, 84–94.

Helgeland, S., Grov, A. and Schleifer, K. H. (1973). *Acta path. microbiol. scand. Sect. B*, **81**, 413–418.

Hobbs, B. C. (1948). *J. Hyg., Camb.*, **46**, 222–238.

Hofstad, T. (1964a). *Acta path. microbiol. scand.*, **61**, 558–570.

Hofstad, T. (1964b). *Acta path. microbiol. scand.*, **62**, 377–380.

Hofstad, T. (1965a). *Acta path. microbiol. scand.*, **63**, 59–71.

Hofstad, T. (1965b). *Acta path. microbiol. scand.*, **63**, 422–434.

Hofstad, T. and Oeding, P. (1962). *Acta path. microbiol. scand., Suppl.*, **154**, 311–312.

Jensen, K. (1958). *Acta path. microbiol. scand.*, **44**, 421–428.

Johnsen, G. S., Endresen, C., Grov, A. and Oeding, P. (1975a). *Acta path. microbiol. scand. Sect. B*, **83**, 226–234.

Johnsen, G. S., Grov, A. and Oeding, P. (1975b). *Acta path. microbiol. scand. Sect. B*, **83**, 235–239.

Juergens, W., Sanderson, A. and Strominger, J. L. (1960). *Bull. Soc. Chim. biol.*, **42**, 110–111.

Julianelle, L. A. and Wieghard, C. W. (1934). *Proc. Soc. exp. Biol. Med.*, **31**, 947–949.

Julianelle, L. A. and Wieghard, C. W. (1935a). *J. exp. Med.*, **62**, 11–21.

Julianelle, L. A. and Wieghard, C. W. (1935b). *J. exp. Med.*, **62**, 31–37.

Karakawa, W. W. and Kane, J. A. (1971). *J. Immun.*, **106**, 900–906.

Karakawa, W. W., Braun, D. G., Lackland, H. and Krause, R. M. (1968). *J. exp. Med.*, **128**, 325–340.

Kikuth, W. and Grün, L. (1957). *Dt. med. Wschr.*, **82**, 549–553.

Kloos, W. E. and Schleifer, K. H. (1975a). *Int. J. syst. Bact.*, **25**, 62–79.

Kloos, W. E. and Schleifer, K. H. (1975b). *J. clin. Microbiol.*, **1**, 82–88.

Kloos, W. E., Tornabene, T. G. and Schleifer, K. H. (1974). *Int. J. syst. Bact.*, **24**, 79–101.

Kocur, M. (1974). *In* "Bergey's Manual of Determinative Bacteriology" (R. E. Buchanan and N. E. Gibbons, Eds.), 8th edn., pp. 489–490. The Williams and Wilkins Company, Baltimore.

Kocur, M., Bergan, T. and Mortensen, N. (1971). *J. gen. Microbiol.*, **69**, 167–183.

Kretzschmar, W. (1969). "Ein einheitliches, natürliches serologisches Klassifizierungssystem für *Staphylococcus aureus* und seine Bewährung in der epidemiologischen Praxis". Thesis, Bautzen, DDR.

Kretzschmar, W. (1976). In "Staphylococci and Staphylococcal Diseases". Proc. 3rd Int. Symp., Warsaw, 1975 (F. Feljaszwics, Ed.), p. 93. Fisher, Stüttgart.

Kretzschmar, W. and Kretzschmar, E. (1962). *J. Hyg. Epidem. Microbiol. Immun.*, **6**, 358–367.

Lachica, R. V. F., Hoperich, P. D. and Genigeorgis, C. (1971). *Appl. Microbiol.*, **21**, 823–826.

Lachowicz, T. (1965). *Zentbl. Bakt. ParasitKde, Abt. I Orig.*, **196**, 340–351.

Lautermann, R. and Grün, L. (1958). *Dt. med. Wschr.*, **83**, 922–924.

Live, I. (1972a). In "The Staphylococci" (J. O. Cohen, Ed.), pp. 443–456. Wiley-Interscience, New York, London, Sydney and Toronto.

Live, I. (1972b). *Am. J. vet. Res.*, **33**, 385–391.

Live, I. and Nichols, A. C. (1965). *J. infect. Dis.*, **115**, 197–204.

Losnegard, N. and Oeding, P. (1963a). *Acta path. microbiol. scand.*, **58**, 482–492.

Losnegard, N. and Oeding, P. (1963b). *Acta path. microbiol. scand.*, **58**, 493–500.

Löfkvist, T. (1957). *Acta path. microbiol. scand.*, **41**, 521–536.

Löfkvist, T. and Sjöquist, J. (1964). *Int. Archs Allergy appl. Immun.*, **24**, 244–253.

Malik, B. S. and Singh, C. M. (1960). *J. infect. Dis.*, **106**, 256–261.

Marandon, J.-L. and Oeding, P. (1967). *Acta path. microbiol. scand.*, **70**, 300–304.

Maskell, R. (1974). *Lancet*, i, 1155–1158.

Mercier, P., Pillet, J. and Chabanier, P. (1950). *Annls Inst. Pasteur*, **78**, 457–466.

Meyer, W. (1967). *Int. J. syst. Bact.*, **17**, 387–389.

Mitchell, R. G. (1965). *Br. med. J.*, **1**, 1127.

Mitchell, R. G. (1968). *J. clin. Path.*, **21**, 93–96.

Mitchell, R. G. and Baird-Parker, A. C. (1967). *J. appl. Bact.*, **30**, 251–254.

Modjadedy, A. and Fleurette, J. (1974). *Annls Microbiol. (Inst. Pasteur)*, **125B**, 367–379.

Moore, E. E. M. (1970). *J. med. Microbiol.*, **3**, 183–184.

Morse, S. I. (1962). *J. exp. Med.*, **116**, 229–245.

Morse, S. I. (1963). *J. exp. Med.*, **117**, 19–26.

Mortensen, N. and Kocur, M. (1967). *Acta path. microbiol. scand.*, **69**, 445–457.

Nathenson, S. G., Ishimoto, N., Anderson, J. S. and Strominger, J. L. (1966). *J. Biol. Chem.*, **241**, 651–658.

Ochi, Y. and Shimizu, T. (1960). *Jap. J. Bact.*, **15**, 713–718.

Oeding, P. (1952a). *Acta path. microbiol. scand.*, **31**, 145–163.

Oeding, P. (1952b). *Acta path. microbiol. scand. Suppl.*, **93**, 356–363.

Oeding, P. (1953). *Acta path. microbiol. scand.*, **33**, 312–323.

Oeding, P. (1954). *Acta path. microbiol. scand.*, **34**, 34–46.

Oeding, P. (1957). *Acta path. microbiol. scand.*, **41**, 310–324.

Oeding, P. (1960). *Bact. Rev.*, **24**, 374–396.

Oeding, P. (1967). *Acta path. microbiol. scand.*, **70**, 120–128.

Oeding, P. (1971). *Int. J. syst. Bact.*, **21**, 323–325.

Oeding, P. (1973). *Acta path. microbiol. scand. Sect. B*, **81**, 327–336.

Oeding, P. (1974). *Ann. N.Y. Acad. Sci.*, **236**, 15–21.

Oeding, P. and Hasselgren, I. L. (1972). *Acta path. microbiol. scand. Sect. B*, **80**, 265–269.

Oeding, P. and Haukenes, G. (1963). *Acta path. microbiol. scand.*, **57**, 438–450.

Oeding, P. and Sompolinsky, D. (1958). *J. infect. Dis.*, **102**, 23–34.
Oeding, P. and Vogelsang, T. M. (1954). *Acta path. microbiol. scand.*, **34**, 47–56.
Oeding, P. and Williams, R. E. O. (1958). *J. Hyg. (Cam.)*, **56**, 445–454.
Oeding, P., Grov, A. and Myklestad, B. (1964). *Acta path. microbiol. scand.*, **62** 117–127.
Oeding, P., Myklestad, B. and Davison, A. L. (1967). *Acta path. microbiol. scand.*, **69**, 458–464.
Oeding, P., Marandon, J.-L., Hájek, V. and Maršálek, E. (1970). *Acta path. microbiol. scand. Sect. B*, **78**, 414–420.
Oeding, P., Marandon, J.-L., Hájek, V. and Maršálek, E. (1971). *Acta path. microbiol. scand. Sect. B*, **79**, 357–364.
Oeding, P., Marandon, J.-L., Meyer, W., Hájek, V. and Maršálek, E. (1972). *Acta path. microbiol. scand. Sect. B*, **80**, 525–533.
Oeding, P., Hájek, V. and Maršálek, E. (1973). *Acta path. microbiol scand. Sect. B*, **81**, 567–570.
Oeding, P., Hájek, V. and Maršálek, E. (1974). *Acta path. microbiol. scand. Sect. B*, **82**, 899–903.
Oeding, P., Hájek, V. and Maršálek, E. (1976). *Acta path. microbiol. scand. Sect. B*, **84**, 61–65.
Parker, M. T. (1972). *In* "Methods in Microbiology" (J. R. Norris and D. W. Ribbons, Eds), Vol. 7B, pp. 2–28. Academic Press, London and New York.
Parker, M. T., Tomlinson, A. J. H. and Williams, R. E. O. (1955). *J. Hyg., (Camb.)*, **53**, 458–473.
Pereira, A. T. (1961). *J. Path. Bact.*, **81**, 151–156.
Peters, G., Pulverer, G. and Pillich, J. (1976). *In* "Staphylococci and Staphylococcal Diseases". Proc. 3rd Int. Symp., Warsaw, 1975 (F. Feljaszewics Ed.), pp. 159–163. Fisher, Stüttgart.
Pillet, J. and Orta, B. (1953). *Annls Inst. Pasteur*, **84**, 420–426.
Pillet, J. and Orta, B. (1954). *Annls Inst. Pasteur*, **86**, 752–758.
Pillet, J. and Orta, B. (1970). *Annls Inst. Pasteur*, **119**, 193–205.
Pillet, J., Isbir, S. and Mercier, P. (1950). *Annls Inst. Pasteur*, **78**, 638–643.
Pillet, J., Mercier, P. and Orta, B. (1951). *Annls Inst. Pasteur*, **81**, 224–227.
Pillet, J., Calmels, J., Orta, B. and Chabanier, B. (1954). *Annls Inst. Pasteur*, **86**, 309–319.
Pillet, J., Rouyer, M. and Orta, B. (1955). *Annls Inst. Pasteur*, **88**, 662–665.
Pillet, J., Orta, B. and Perrier, M. (1961). *Annls Inst. Pasteur*, **101**, 590–595.
Pillet, J., Orta, B., Perrier, M. and Corrieras, F. (1962). *Annls Inst. Pasteur*, **103**, 716–727.
Pillet, J., Orta, B., Corrieras, F. and Perrier, M. (1966). *Annls Inst. Pasteur*, **110**, 422–435.
Pillet, J., Orta, B. and Corrieras, F. (1967). *Annls Inst. Pasteur*, **113**, 363–374.
Pillet, J., Orta, B., Corrieras, F. and Petillon, C. (1968). *Annls Inst. Pasteur*, **114**, 658–668.
Plommet, M. G. and Wilson, J. B. (1969). *J. comp. Path. Ther.*, **79**, 425–433.
Pulverer, G. (1965). *Fortschr. Med.*, **83**, 459–462.
Quinn, E. L., Cox, F. and Fisher, N. (1965). *Ann. N.Y. Acad. Sci.*, **128**, 428–442.
Reiss, J., Lachowicz, T. and Lacki, W. (1969). *Arch. Immun. Therap. Exp.*, **17**, 153–158.
Richmond, M. H. (1972). *In* "The Staphylococci" (J. O. Cohen, Ed.), pp. 159–186. Wiley–Interscience, New York, London, Sydney and Toronto.

Roberts, A. P. (1967). *J. clin. Path.*, **20**, 631–632.

Sanderson, A. R., Juergens, W. G. and Strominger, J. L. (1961). *Biochem. biophys. Res. Commun.*, **5**, 472–476.

Schindler, C. A. and Schuhardt, V. T. (1964). *Proc. natn. Acad. Sci. U.S.A.*, **51**, 414–421.

Schleifer, K. H. and Kandler, O. (1970). *J. Bact.*, **103**, 387–392.

Schleifer, K. H. and Kandler, O. (1972). *Bact. Rev.*, **36**, 407–477.

Schleifer, K. H. and Kloos, W. E. (1975a). *Int. J. syst. Bact.*, **25**, 50–61.

Schleifer, K. H. and Kloos, W. E. (1975b). *J. clin. Microbiol.*, **1**, 337–338.

Sekiya, C. (1970). *Acta paediat. jap.*, **12**, 71–73.

Shimizu, T. (1968). *Mem. Fac. Agric. Miyazaki Univ.*, **5**, 1–55.

Shulman, J. A. and Nahmias, A. J. (1972). *In* "The Staphylococci" (J. O. Cohen, Ed.), pp. 457–481. Wiley–Interscience, New York, London, Sydney and Toronto.

Slanetz, L. W. and Bartley, C. H. (1962). *J. infect. Dis.*, **110**, 238–245.

Smith, H. B. H. and Farkas-Himsley, H. (1969). *Can. J. Microbiol.*, **15**, 879–890.

Smith, I. M., Beals, P. D., Kingsbury, K. R. and Hasenclever, H. F. (1958). *Archs intern. Med.*, **102**, 375–388.

Sompolinsky, D., Hermann, Z., Oeding, P. and Rippon, J. E. (1957). *J. infect. Dis.*, **100**, 1–11.

Stern, H. and Elek, S. D. (1957). *J. Path. Bact.*, **73**, 473–483.

Subcommittee on Phage Typing of Staphylococci (1971). *Int. J. syst. Bact.*, **21**, 167–170.

Subcommittee on Phage Typing of Staphylococci (1975). *Int. J. syst. Bact.*, **25**, 233–234.

Subcommittee on Taxonomy of Staphylococci and Micrococci (1965). *Int. Bull. bact. Nomencl.*, **15**, 107–110.

Subcommittee on Taxonomy of Staphylococci and Micrococci (1976). *Int. J. syst. Bact.* **26**, 333–334.

Verhoef, J., van Boven, C. P. A. and Winkler, K. C. (1972). *J. med. Microbiol.*, **5**, 9–19.

Verwey, W. F. (1940). *J. exp. Med.*, **71**, 635–644.

Vischer, D. (1959). *Schweiz. Z. allg. Path. Bakt.*, **22**, 42–62.

Wahl, R. and Fouace, J. (1952). *Annls Inst. Pasteur*, **82**, 542–555.

Wiley, B. B. (1972). *In* "The Staphylococci" (J. O. Cohen, Ed.), pp. 41–63. Wiley–Interscience, New York, London, Sydney and Toronto.

Yoshida, A., Mudd, S. and Lenhart, N. A. (1963). *J. Immun.*, **91**, 777–782.

Yoshimura, H. (1970). *Jap. J. vet. Sci.*, **32**, 263–274.

CHAPTER VIII

# Group and Type (Groups A and B) Identification of Haemolytic Streptococci

## J. ROTTA

*Institute of Hygiene and Epidemiology, Prague*

## I. INTRODUCTION

Although considerable improvements have been made in the diagnosis, therapy and prevention of streptococcal infections in the moderate climatic zone, these diseases still represent a public health problem which cannot be disregarded. The data on the incidence of streptococcal infections in sub-tropical and tropical areas, as collected in recent years, clearly document that the role of streptococci in these parts of the world has been seriously underestimated. Streptococcal infections are of worldwide importance.

Because the diagnosis of streptococcal infection based only on clinical examination is never certain, microbiological examination of all cases suspected of streptococcal etiology represents an important component of the recognition and identification of the disease. Such an approach enables the institution of early and adequate treatment. Since the sequelae of streptococcal infections, such as rheumatic fever and acute glomerulo-nephritis, have conclusively been recognized to follow only after infection caused by group A streptococci, the direct or indirect identification of antecedent infection by this agent is of basic importance for establishing the diagnosis of these conditions.

The microbiological procedures available for the recognition of haemo-lytic streptococci in the specimen to be examined and identification of the serological group of the isolated streptococcus belong among the basic laboratory techniques. They should be employed at all levels of micro-biological laboratory services. The examinations are simple and inexpen-sive. However, they must be carried out in an appropriate way in order to provide a reliable result. Type classification of strains belonging to group A and B streptococci are specialised procedures and are executed mainly in laboratories on the central level of health services.

## II. BASIC BIOLOGICAL CHARACTERISTICS OF HAEMOLYTIC STREPTOCOCCI

A large number of biologically diverse micro-organisms belong to the genus *Streptococcus*. They are Gram-positive organisms, spherical or ovoid in shape, growing in pairs or short or long chains. While some species are primary pathogens for man and animals, others can produce pathologic processes only under particular conditions or are purely saprophytic, forming part of the normal microbial flora in some organs of the human or animal body.

Various schemes for the classification of streptococci have been elabora-ted, but no fully satisfactory system has yet been established. The available schemes usually give preference to one or a few biological characteristics while others are not considered to be of the same importance. However, the taxonomic approach most commonly used in identifying haemolytic streptococci in bacteriological practice at present has many advantages. It is based on two major characteristics, haemolysis and the presence of a polysaccharide antigen in the cell wall. This approach is very advantageous because it enables us to recognise most of the streptococcus strains primarily or conditionally pathogenic for man and animal.

### A. Definition of haemolytic streptococci

Haemolytic streptococci easily grow on blood agar containing various mammalian red blood cells and cause their lysis. The majority of strains of most serological groups produce clear and complete haemolysis (beta haemolysis) on the blood agar. However, with some strains of particular groups (e.g. groups B, D, K, O) the haemolysis takes the form of greening (alpha haemolysis) or no haemolysis is produced.

The property of primary importance in identifying the haemolytic streptococcus is the presence of a polysaccharide antigen in the cell wall or of a teichoic acid antigen in the cytoplasm. Both substances are serologically distinct and differentiate haemolytic streptococci into definite serological groups.

Among the haemolytic streptococci of groups A to V that have been described so far the enterococci represent micro-organisms which differ in several biological respects from the streptococci of other groups. Entero-cocci are resistant to a temperature of 60°C for 30 min, they grow at pH 9·6 and in the presence of 6·5% NaCl. Group D streptococci include three species of enterococci (*S. faecalis*, *S. faecium*, *S. durans*) and two species of "enterococcus-like streptococci" (*S. equinus*, *S. bovis*). The identification scheme for the species within group D streptococci is based on physiological and biochemical tests. A phage typing system has also been elaborated. The subdivision of groups other than A, B, and D has been undertaken in particular groups, e.g. groups C, G, N. However, the systems worked out so far do not represent a considerable contribution to practice and are employed rather exceptionally.

## B. Growth requirements and colony forms

Although considerable variation exists between the particular species of the genus *Streptococcus* in nutritional requirements, all streptococci require media rich in nutrients. This is essential in order to achieve satisfactory growth of cells with an adequate antigenic equipment.

The basic culture medium must contain meat infusion (freshly prepared is preferable to commercial dehydrated), peptone, glucose and salts. Sterilisation by heat reduces the quality of the medium in respect of growth. However, heat sterilisation is fully satisfactory for most purposes and is generally used. For delicate cultivation purposes filtration of the medium is superior to sterilisation by heating. The addition of serum or blood into the medium greatly enhances the growth of streptococci. Such enrichment is especially necessary for cultivating streptococci on solid media (blood agar plates), but is also recommended for the pre-paration of highly virulent cultures grown for a short period of time (serum broth).

Studies on the growth of streptococci under defined cultivation con-ditions have provided information on their minimal nutritional requirements. These data do not have a direct importance for routine diagnostic laboratory work but are essential for a better understanding of the physiological mechanism operating during growth. For example, good growth of streptococci was obtained on a chemically defined medium containing 22 amino acids, six vitamins, some purines and pyrimidines (Ginsburg and Grossowicz, 1957). However, the most essential growth factor(s) for streptococci seems to be a peptide which has been named "strepogenin" by Wooley (1941). The lack of satisfactory growth on most chemically defined media is due to the absence of this substance. Strepogenin can be partly substituted by glutamine and asparagine (Kodíček and Mistry, 1955). Since strepogenin has mostly been studied in streptococci, the substance is sometimes designated "streptogenin".

Haemolytic streptococci utilise polysaccharide as an energy source. The resulting metabolic product is lactic acid. Haemolytic streptococci are facultative anaerobes. The acidic pH developing during the production of lactic acid produces inhibition of further growth. In order to achieve more vigorous growth, continuous neutralisation of the culture medium is essential. However, this is not necessary in routine work when cultures of streptococci for grouping and typing are prepared.

Similarly as with most other bacteria pathogenic for man, the optimal temperature for the growth of streptococci is 37°C. Heating to 56°C reliably kills most strains of haemolytic streptococci but some species survive this temperature (e.g. the enterococci).

Haemolytic streptococci readily grow on blood agar and produce complete lysis of red blood cells around the colonies. There is great variation in the shape and size of the lysed zones. This depends not only on the lytic properties of the particular streptococcus but also on the kind of blood used and on the cultivation conditions. Moreover, some strains belonging to particular groups (e.g. groups B, D, H, O) produce greening (alpha) haemolysis or are non-haemolytic. Haemolytic streptococci grow in three main colony forms on blood agar: **mucoid** colonies, which are large and watery, with a raised surface—streptococci growing in this phase regularly produce hyaluronic acid capsules and frequently the M protein; **matt** colonies, which are opaque and flat, with a tendency to produce a rough surface because in most instances they develop from mucoid colonies as a result of drying and for this reason are sometimes called postmucoid colonies—M protein is frequently produced; **glossy** colonies, which are smooth and glistening in appearance, with a diameter much smaller than in the other colony forms, usually 1–2 mm only—only seldom is the M protein present. It should be noted, however, that the relation between colony form and M protein production is approximate only and should never be considered a definite characteristic because there is considerable variation in these two properties of haemolytic streptococci.

Haemolysis and colony forms are the main characteristics of haemolytic streptococci for their identification on blood agar plates. It is essential to distinguish them from haemophili and pyogenic corynebacteria because of the resemblance of the haemolysis. In the case of uncertainty, a Gram-stained smear should be examined.

## C. Structure and antigenic composition of surface layers

The surface layers of haemolytic streptococci have been recognised to be the most important components in respect of the classification of these pathogens.

The **capsule** is the outermost layer of the cell. Its production depends on the properties of the particular strain and on the cultivation conditions. In group A and C streptococci, for example, the capsule is composed of hyaluronic acid consisting of N-acetylglucosamine and glucuronic acid.† Strains producing a relatively large quantity of the capsular material grow in mucoid colonies on blood agar plates. It has been found that group A streptococci growing with such colonies contain M protein (virulence factor) more frequently than strains growing in other colony forms. Therefore the identification of mucoid colonies has a definite value in diagnostic work as far as the virulence of group A streptococcus is concerned.

The hyaluronic acid of group A and C streptococci is not antigenic for man and animals and of no importance for classification.

In group B streptococci the capsule is made of polysaccharide, which represents the type-specific substance here. Its chemical and serological specificity enables recognition of types Ia, Ib, II and III. The new protein antigen recently described makes it possible to distinguish the new type Ic.

The **cell wall** of haemolytic streptococci consists of peptidoglycan, which ensures the rigidity of the cell. It has various biological properties, but has no importance for classification because of its similarity in various species of the genus *Streptococcus*. However, the other two major components located in the cell wall, namely the polysaccharide and the proteins, are of considerable importance for classification.

The **polysaccharide** of the cell wall is group-specific and on the basis of its serological specificity haemolytic streptococci can be differentiated into groups A–V. The determinants of polysaccharides of some groups have been identified. For example, in group A, the determinant is N-acetylglucosamine, in group C it is N-acetylgalactosamine. The molecule

† Some strains of group D streptococci may also produce capsules composed of hyaluronic acid.

of the polysaccharide of group A streptococcus is multibranched, containing 17 moles of N-acetylglucosamine and 38 moles of rhamnose; 11 moles of N-acetylglucosamine are at the end position of the side chains of the molecule and can be split-off by N-acetylglucosaminidase. Further studies have determined the nature of attachment of the polysaccharide to the peptidoglycan (Haverkorn, 1973).

The **M protein** is the type-specific antigen used for the differentiation of 65 types within the group A streptococcus. Identification of the M protein in group A streptococci is also important because of its role as a virulence factor. Antibody against the M protein develops in streptococcal infections and produces type-specific immunity.

The M protein was recognised earlier than the group-specific substance. It was originally anticipated that all beta-haemolytic streptococci isolated from man belong to one group. This has later turned out to be erroneous. As a result several type numbers are not used any more, e.g. the original types 7, 20 and 21, which have been shown to carry the group C polysaccharide, and type 16, belonging to group G. Moreover, several candidates for new types were not previously sufficiently analysed and it was later found that they actually belong to M types already recognised. For example, types 10 and 12 are identical and so are types 35 and 49. Only the numbers 12 and 49 are used for these types now.

Another important finding was made when highly purified M proteins of various types were studied antigenically. It has turned out that antigenic relationship exists between types 13 and 48; 2 and 48; 3 and 12; 33, 41, 43 and 52. The serological cross-reactivity is not seen in routine typing. However, it must be anticipated that it might play some role in streptococcal immunity (Fox and Wittner, 1968; Willey and Bruno, 1968).

Although no prevalence of particular serological types in streptococcal diseases of the upper respiratory tract of man has been found, studies on streptococcal skin infections have revealed a definite affinity of certain types to the skin. These are mainly the types above type 50. Furthermore, certain types have evident nephritogenic properties and acute glomerulonephritis is seen after infection with types 4, 12, 25, 49 and several types above 50 more frequently than after infection provoked by other types. The search for rheumatogenic types has failed so far.

Extensive studies have been carried out in recent years on the location of the M protein in the wall and its production, isolation and purification. It was shown that types 1, 6, 12, 14 and 28 (Swanson *et al.*, 1969) carry the M protein in fimbriae covering the surface of the cell wall. The fimbriae are attached to the peptidoglycan layer and each of the fimbria is composed of several segments containing the M protein. A description of the isolation and purification of M protein lies beyond the scope of this Chapter. It should only be noted that for routine extraction of M protein for typing purposes, the cells are heated to 98 °C at pH 2·0. This extraction is the first step of several procedures for preparing highly purified materials.

Only fresh group A strains isolated from patients or strains properly stored should be used for cultivation for typing purposes. The number of passages on artificial media before typing should be kept to a minimum. The media must contain peptides, glucose, and reducing agents (Davies *et al.*, 1968).

Amino-acid analysis of various preparations of M protein revealed data of limited value. The figures are not comparable because of great differences in the purity of the materials used in these studies.

Research on the antigenicity and chemistry of M proteins is essential for work aimed at the elaboration of streptococcal vaccine for use in humans. It is considered that the vaccine would be of some importance in selected population groups or individuals. A prerequisite for its employment is knowledge of the prevalence of particular types of group A streptococcus in various geographical areas over a long period of time.

There is an immunological relationship of streptococcal antigens to mammalian heart tissue components and this is of significance for the study of the pathogenesis of streptococcal infections and their sequelae (Lyampert *et al.*, 1966; Kaplan, 1963). These streptococcal antigens are related to the M protein and to the components of the cytoplasmic membrane.

The **M-associated protein (MAP)** has been identified relatively recently. It is a non-type-specific substance, but closely associated with the M protein (Widdowson *et al.*, 1971). M protein negative strains do not contain this protein. Although the M-associated protein has no importance for type classification of group A strains, the finding of this antigen in a group A strain suggests that the strain carries the M protein.

However, M protein and M-associated protein may represent two parts of one molecule and further study is needed to clarify their relationship. The **serum opacity factor (SOF)** was found to be a second type-specific substance in some types of group A streptococci. These streptococcus types are those which produce poorly antigenic M protein, against which M typing sera are difficult to prepare. To date, the SOF has been demonstrated in 16 types. Good anti-SOF rabbit sera can be prepared and used for classifying these types (Widdowson *et al.*, 1970). These typing sera include, for example, antisera against types 4 and 22, which can very rarely be typed by the M protein. There still remains much to be studied concerning the SOF before a conclusive assessment of its role for the streptococcus and for the host can be drawn.

The **T protein** is the cell-wall antigen, predominantly present in group

A streptococci, which forms the basis for typing by the agglutination reaction. The T protein does not play any role in the virulence of the streptococcus.

Serologically distinct T proteins are usually shared by two or more M types. Moreover, M types possess more than one antigen as a rule. Only exceptionally one T protein is type-specific for one M type, as is the case with types 1, 6, 9, 18 and 22. The term "T pattern" has been introduced for "types" determined according to the T protein (Stewart et al., 1944; McLean, 1953). The following symbols are used at present for T patterns (in parentheses are the M types which each T pattern includes): $3/13/B_{3264}$ (3, 13, 33, 39, 41, 43, 52, 53, 56); 4/28 (4, 24, 26, 28, 29, 46, 48, 60); 5/12/27 (5, 11, 12, 27, 44, 61); 14/49 (14, 49); 15/23/47 (15, 17, 19, 23, 30, 47, 54); 8/25/Imp 19 (2, 8, 25, 31, 55, 57, 59). The T antigenic structure has not yet been determined in all M types and it is therefore clear that the above scheme is not definitive.

Group C and G strains may contain protein antigens serologically identical with some of the T proteins of group A streptococci, for example the T antigens 2, 4, 8, and 25.

No information is available on the chemical composition of T protein because no work on the purification of this component has yet been carried out.

**R protein** is a cell-wall component present in some strains of particular group A types and in some strains belonging to groups B, C and G (Lancefield, 1943, 1957; Maxted, 1949). This antigen has no relation to virulence or immunity.

The R protein occurs in two serological variants: R 28, first described in type M 28, and R 3 found in type M 3. While the R 28 protein occurs in a number of types and groups, the R 3 protein is only present in type M 3 strains (Lancefield, 1958).

The **peptidoglycan** (other terms used are mucopeptide and murein) which provides the rigidity of the cell is the basic structural component of the cell wall. The molecule of streptococcal peptidoglycan is well-defined. It consists of repeating units of N-acetylglucosamine and N-acetylmuramic acid. The peptides composed of alanine, glutamic acid and lysine in a ratio of 3:1:1 are attached to the muramic acid through its carboxyl group. The alanine dipeptide bridges, which form the cross-linkage of tetrapeptides to adjacent hexosamine polymers, are composed of L-alanyl-L-alanine. They join the terminal alanine of one peptide to the epsilon-amino-group of lysine of the adjacent peptide (Caravano, 1968; Petit et al., 1966).

Peptidoglycan has no importance for differentiating between the various species of the genus *Streptococcus* However, its minor antigenic differences in substances derived from various bacteria indicate that this component might be used in the taxonomy of bacteria in the future.

Peptidoglycan has a variety of pronounced biological activities, such as pyrogenicity, capability of producing the local Shwartzman reaction, lytic effect on blood platelets, etc.

The **polyglycerophosphate** of the *Streptococcus* is chemically a teichoic acid (McCarty, 1959). In the cell, one molecule of alanine is bound to the polyglycerophosphate molecule, but the alanine rapidly splits off in extracted material kept at a temperature above $0\,°C$. Both the polyglycerophosphatealanine and polyglycerophosphate have different serological specificities. So far, no use has been found for the component in classification. The polyglycerophosphate is a part of the lipoteichoic acid molecule located in fimbriae, enabling streptococci to adhere to human cells.

## III. GROUPING OF HAEMOLYTIC STREPTOCOCCI

### A. Importance of group identification

In man, nearly all cases of acute respiratory streptococcal infections are caused by the group A streptococcus. This streptococcus is also responsible for almost all infections spreading as epidemics. Group B streptococci frequently cause disorders in the urogenital tract of women. Skin lesions in man are in a high proportion produced by group A streptococci, but group C and G streptococci are more frequently isolated in this clinical pattern than in respiratory streptococcal disease. The sequelae of acute streptococcal infections, such as acute rheumatic fever and acute glomerulonephritis, only follow after infections produced by group A streptococci. Carriership of beta-haemolytic streptococci in the upper respiratory tract is, generally speaking, very frequent. Microbiological examinations show that haemolytic streptococci of groups other than A are more often responsible for carriership than for disease.

Group identification of streptococci is therefore indispensable in clinical microbiological and epidemiological work. Today it should belong to the routine procedures of every microbiological diagnostic laboratory.

### B. Methods currently used, their principles and suitability for laboratory work

At present many different ways of serological grouping are used; screening tests for the direct identification of groups A and B have also been elaborated. In serological group identification by means of the precipitin reaction, the reagents are hyperimmune rabbit grouping serum and the group antigen of the examined strain prepared by extraction according to one of the following methods: with hydrochloric acid (Lancefield, 1933), with formamide (Fuller, 1938), by liberation with phage lysin (Maxted, 1953), by enzyme from *Streptomyces albus* (Maxted, 1948) or even by simple autoclaving (Rantz and Randall, 1955). It is likewise possible to use the method of precipitation in agar (Lancester and Sherris, 1960; Michael and Massel, 1965; Kunter, 1963) or precipitation on cellulose membrane (Goldin and Glenn, 1964). The fluorescent antibody method is now also widely used (Moody *et al.*, 1958; Wagner and Heinrich, 1962; Redys *et al.*, 1963).

Grouping sera are commercially available. They can be easily produced in every laboratory which has the basic laboratory equipment and has the facilities to keep rabbits for immunisation. The information on grouping sera production may be derived from a number of papers and is also contained in bulk in a methodical laboratory pamphlet (which can be supplied on request, Rotta, 1970).

Of screening tests based on physiological reactions, one should mention the Bacitracin test of Maxted (1953) and the triple test of Wallerström (Wallerström, 1962), for group A identification, and the CAMP test for group B identification (Munch-Petersen and Christie, 1947).

The Bacitracin test is employed in many laboratories as a screening procedure for group A streptococci, which are highly sensitive to this antibiotic. A decisive factor in this test and its practical applicability is the amount of Bacitracin contained in the paper discs. Some group C, G and L strains display Bacitracin sensitivity approximating to that of some less sensitive group A strains and therefore cannot be differentiated from these by this method.

The CAMP test is based on the interference between staphylococcus beta toxin and group B streptococcus haemolysin. It should be pointed out, however, that the phenomenon is not strictly specific for group B streptococci, for it is also encountered with some group E, P, U and V strains.

The serological method of group identification is the most appropriate and reliable procedure. The methods of extracting the group antigen from the cells according to Lancefield and Fuller are the most frequently used procedures because of their full suitability. They will be described in the following text.

## C. Serological grouping by the precipitation reaction

1. *Preparation of grouping extract*

(a) *Fuller's method* (Fuller, 1938). Add 0·1 ml formamide ($H.CONH_2$) to bacteria harvested by centrifugation of 5 ml 16-h culture of a streptococcus strain grown in Todd-Hewitt broth. Shake thoroughly and heat in an oil bath at 150 °C for 10 min. Cool and add 0·25 ml acid alcohol (composition: 1 ml conc. HCl in 99 ml of 95% $C_2H_5OH$); shake; separate the precipitate by centrifugation and add 0·50 ml acetone to the clear supernatant fluid. A polysaccharide precipitate will form, which is separated by centrifuging and dissolved in 0·35 ml sterile saline. Add one drop of Phenol red solution as indicator (preparation of the solution: dissolve 0·1 g Phenol red in 28 ml 0·01 N NaOH and adjust volume to 250 ml with $H_2O$), and neutralise with N/5 NaOH.

(b) *Lancefield's method* (Lancefield, 1933). Add 0·50 ml N/5 hydrochloric acid to bacteria harvested by centrifugation of 50 ml of a 16 h streptococcus culture grown in Todd-Hewitt broth, shake thoroughly and place the suspension in a boiling water bath for 10 min. After cooling the suspension, add one drop of Phenol red solution as indicator (composition as in (a) above), and neutralise at first with 4 N NaOH and then with N/5 NaOH. Centrifuge; the supernatant fluid represents the extract containing the group polysaccharide.

*Note:* with group M strains, cultivation in broth with serum or on serum agar is preferable. With group D strains (enterococci) cultivation in dextrose broth (pH 7·2) without phosphates is recommended.

## 2. *Performance of the test for group identification*

When using Fuller's extract, the group identification is carried out in special conically narrowing capillaries, fused at one end. The external diameter of the wider part is approximately 5 mm, the total length 3–4 cm. The capillary is fixed in a rack with its sharp point stuck in a groove filled with plasticine. Use a Pasteur pipette first to deliver serum to reach to the middle of the narrowing part. Then deliver extract taking care that the two substances do not mix or an air bubble does not form between them. Read the formation of the precipitin ring at the place of contact of the fluids against a dark background within 5 min.

When using Lancefield's extract, the grouping test is performed in capillary tubes of an external diameter about 1 mm and length about 8 cm. The capillary is applied in a slanting position to the surface of the serum and a column of about 2 cm is drawn in; the tip is wiped with cellulose cotton wool and applied to the extract surface, when 2 cm of extract are drawn in a similar manner. The column of the two fluids is drawn up so as to occupy the middle of the capillary and the capillary is inserted in a vertical position in a rack with a groove filled with plastic clay. The precipitation reaction should occur within 5 min at room temperature.

3. *Evaluation of results*

In clear-cut reactions, an evident precipitate should form a ring or fill the capillary tube.

Reactions taking longer than 5 min must not be evaluated as positive.

The method giving the most reliable results is that of Fuller: it produces fewest cross reactions. However, it has certain disadvantages: it is rather time-consuming and the high extraction temperature may destroy the group polysaccharide, e.g. in group O and N.

Lancefield's method is in most instances a satisfactory procedure in laboratory practice.

Dubious or non-specific precipitation reactions may be due to either alkaline extracts (pH higher than 7·6) or contaminated, sometimes opalescent sera and extracts.

A very important factor in streptococcus grouping is the use of rabbit sera with a sufficient titre of group-specific antibody. It has been shown that for serological identification of group A streptococci by Lancefield's extract the rabbit serum must contain at least 5 mg of immunoglobulin in 1 ml of serum. However, sera containing less immunoglobulin may give satisfactory reactions when using Fuller's extract in the ring test.

## IV. TYPING OF GROUP A STREPTOCOCCI

### A. Importance of type identification

In the majority of cases of acute streptococcal infections examined in routine practice, the type identification of group A streptococci is not a necessary procedure. However, in particular cases of streptococcal disease, and in sequelae, information on the type of the streptococcus involved is very valuable. This especially applies to the spread of nephritogenic types and of types which are known to be of particular virulence in a geographical locality at a given time. The main importance of streptococcus typing is in epidemiological tracing of streptococcus circulation among the population and in microbiological and epidemiological studies of streptococcal infections and their sequelae. These studies supply data on the distribution of streptococcus types and on the dynamics of changes in this distribution. This is of major importance for the prospect of using streptococcal vaccine in the future.

### B. Methods currently used, their principles and suitability for laboratory work

Since the early days of streptococcus bacteriology, two systems have been used in typing, namely Lancefield's system (Lancefield, 1928) based on the identification of M types by the precipitation test and Griffith's system (Griffith, 1934) based on the identification of T-patterns by the agglutination test. These systems have gradually been greatly improved and refined so that they represent very valuable typing procedures at present.

Typing sera are not commercially available, except T typing sera, which are supplied by one firm only. The production of M and T typing sera requires good laboratory facilities and sufficient space for rabbits. The procedures how to produce and check the sera have been described by several authors, e.g. Moody et al. (1965), Facklam and Moody (1968), Rotta (1975), Rotta et al. (1971).

The M typing procedure should be given priority over T typing, although usually only less than 50% of group A strains collected from various sources are classifiable according to the M protein and more than 90% of these strains are identifiable by their T protein. The rest of the strains are not M typable, because a high number of group A streptococci lose the capability to produce the M protein when subcultured on bacteriological media. Furthermore, the rabbit hyperimmune typing sera frequently contain insufficient titre of anti-M antibody or the typing serum is not available for the strain to be typed.

The reasons for the priority of M typing are as follows: first, the M protein, as the only one type-specific substance in group A streptococcus determines the type most precisely; second, the M protein has a special importance because it is a factor of virulence.

T typing should be carried out in parallel with M typing, although it supplies in the majority of cases only gross information on the type of the strain. However, in many instances the T typing will give information on the type identity or difference of two or more strains.

## C. M typing

For decades, M typing has been performed as the capillary precipitin test described by Swift et al. (1943) using absorbed type-specific anti-M sera and hydrochloric acid extracts of the strain being examined. A method recently developed by Rotta et al. (1971) uses unabsorbed anti-M sera in a double diffusion test. It is comparable with the capillary test in terms of result, but is simpler, faster and easier to perform. This method is especially advisable in newly established streptococcus laboratories without much experience in streptococcus bacteriology.

### 1. M typing with absorbed sera

Hyperimmune rabbit sera containing a sufficient level of anti-M protein precipitating antibody should be used for this typing procedure. The sera should be absorbed in order to eliminate antibodies to antigens common to group A streptococci, mainly to the group-specific polysaccharide.

The test is carried out as a precipitation reaction in capillary tubes. Two components form the reacting system: the M protein extract from the group A streptococcus to be typed and anti-M absorbed serum. The method is time consuming and only potent absorbed typing sera must be used in order to achieve reliable results. The reaching of this goal is in many instances not without difficulties.

(a) *Preparation of typing extract.* Add 0·40 ml of N/5 hydrochloric acid to bacteria harvested by centrifugation from 80 ml streptococcus culture grown for 16 h in Todd-Hewitt broth, shake thoroughly and place the suspension in a boiling water bath for 10 min. After cooling it, add one drop of Phenol red solution as indicator (preparation of the solution: dissolve 0·1 g of Phenol red in 28 ml of 0·01 N NaOH and adjust volume to 250 ml with $H_2O$) and neutralise at first with 1 N NaOH and then with N/5 NaOH. Spin down, the supernatant fluid represents the extract containing the M protein.

*Important note:* it is essential that the extraction of M protein should be performed at pH 2; in a more acidic medium hydrolysis and destruction of M protein occurs. For this reason, it is recommended that the pH of the streptococcus suspension in HCl be checked before boiling by means of Thymol blue solution (preparation of the solution: dissolve 0·1 g of Thymol blue in 4·3 ml of 0·05 N NaOH and adjust the volume to 100 ml with distilled water). One drop of the suspension is mixed with one drop of the indicator on a slide; the colour must be only light pink, red colouration indicating too low a pH.

In cases of poor M typability it may prove advantageous to increase the content of Neopeptone in the Todd-Hewitt broth by adding 10 or 20 g of Neopeptone per 1000 ml in Todd-Hewitt broth.

(b) *Performance of the precipitation reaction.* The sera used for M typing are not commercially available. They are produced in several national reference laboratories in various parts of the world.

The precipitation reaction is carried out in capillary tubes. The tubes should have an external diameter of about 1 mm and should be about 80 mm long. They must be adequately washed and dried.

Suck into the capillary tube a column of about 2 cm by applying the tube in a slanting position to the surface of the serum. By using a cellulose cotton-wool, clean the end of the capillary tube and suck-up, in a similar way a column of about 2 cm of extract by applying the same end of the capillary tube to the surface of extract. No bubble should develop between the two fluids in the capillary tube. Fix the capillary in the plasticine by the same end in a vertical position.

Set up the test with all anti-M sera available, each serum in a different tube. If the T typing is carried out parallel with the M typing, the number of M sera can be restricted according to the result of the T typing.

The recommended selection of sera for M typing is indicated in the following scheme:

| T pattern identified | M sera to be used in capillary test† |
|---|---|
| 1 | 1, 3 |
| 3/13/B$_{3264}$ | 3, 13, 33, 39, 41, 43, 52, 53, 56, 3 R, 1, 12 |
| 4/28 | 4, 24, 26, 29, 46, 48, 60, 28 R |
| 5/12/27 | 5, 11, 12, 27, 44, 61, 1, 3 |
| 6 | 6, 1, 2, 12 |
| 8/25/Imp 19 | 2, 8, 25, 31, 55, 57, 59, 1, 6 |
| 9 | 9, 1 |
| 14/49 | 14, 49, 51, 12, 1 |
| 15/23/47 | 15, 17, 19, 23, 30, 47, 54, 18 |
| 18 | 18, 1, 15, 17, 19, 23, 30, 47, 54 |
| 22 | 22, 1, 12 |

† Sera of several unrelated types are included for checking the specificity of the precipitation reaction.

Such a selection is possible since it is known that the so called T patterns include only a limited number of M types.

Place the rack with tubes in a thermostat at 37°C for 2 h. Read the precipitation reactions (see the scale later) and record the results. Transfer the rack with the tubes to a refrigerator at +4°C overnight and read the reactions again.

(c) *Evaluation of the results.* The evaluation is made according to the following scale:

+,    evident precipitation, the height of the precipitation column amounting to 2 mm maximum,

++,    the precipitate column amounts to 3–10 mm,

+++,    more than 10 mm of the capillary is filled with precipitate.

Record only clear and evident precipitation reactions. In general, the precipitation reaction should develop between the extract and one serum only. Such a result clearly identifies the M type. If a precipitation reaction develops with two or more sera, the typing should be repeated using a new extract. M-absorbed sera should be checked with M extracts of reference strains if there is any uncertainty about the specificity of the sera.

Strains which do not possess the M protein at the time of typing cannot be type identified by this method. In current laboratory practice a large proportion of strains may lack this protein pattern. It is recommended that freshly isolated strains be typed; passage and subculture in artificial media reduce the production of the M protein.

## 2. *M typing with unabsorbed sera*

Typing is carried out by the method of Rotta *et al.* (1971).

This procedure is carried out as a double-diffusion test in agar gel, it is simpler, faster and easier to perform than the precipitin test in capillary tubes. Both methods reveal very good agreement in routine practice.

The identification of precipitation bands corresponding to the group and type reaction is possible because the bands develop in different positions due to different diffusion patterns of the M protein and of the group A polysaccharide. However, some unabsorbed sera may give cross-reactions by bands mimicking the M/anti-M reaction, but this is usually only the property of a particular serum.

Cross-reaction is not seen if an anti-M serum produced in another rabbit is used. The unabsorbed sera should therefore be checked before they are used in agar gel.

(a) *Preparation of typing extract.* Use the same procedure as indicated in paragraph (a), page 190.

(b) *Performance of the precipitation reaction in agar gel.* Sera for M typing in agar gel are not commercially available. They can be requested from national streptococcus reference laboratories where such sera are produced.

Prepare a 1% solution of Noble agar in 0·01 M phosphate buffer saline, pH 7, and add sodium azide to a final concentration 0·01%. Melt well by steaming the bottle for 15 min and transfer it into a water bath of 50 °C. For one test a volume of about 100 ml of the agar solution is needed. It is convenient to prepare a stock amount in a number of bottles, store them in a refrigerator at +4 °C and melt the agar before use in a boiling water bath. Pour the agar solution (heated to 50 °C) into four plastic frames holding six glass microscope slides each. About 25 ml of agar solution is needed for each frame. After solidification, cut out over each slide two sets of wells, each consisting of eight peripheral wells and one central well. The distance of the centres of each peripheral well to the central well is 7·5 mm. The diameter of the wells is 3 mm. The plastic frames and the punch for cutting the wells are manufactured by Gelman Instrument Co., Ann Arbor, Mich., USA.

Using Pasteur pipettes, fill the central well with serum and the peripheral wells with extracts of the strains to be typed. Thus the number of sets of wells corresponds to the number of sera available for typing. Eight strains can be typed at a time against one serum. It is not possible to place the extract into the central well and the sera into the peripheral wells, because of frequent cross-reactions due to the large volume of serum in the plate. Incubate in a moist chamber at room temperature (22 °C approximately) overnight.

(c) *Evaluation of the results.* The band corresponding to the M/anti-M reaction develops nearer to the well containing the extract while the band representing the group reaction forms close to the central well. The intensity of the bands reflects the potency of the sera and the quantity of the M protein in the extract. Cross-reactions mimicking the type reaction occur very rarely. Extraction of the streptococcus cell at pH 2 destroys a variety of cell proteins including the T antigen.

The M type is identified if the extract of the strain gives a clear precipitation reaction with one serum only. If cross-reactions are seen, the typing in agar gel should be repeated with a new hydrochloric acid extract.

Similarly as in M typing by the capillary test, a positive result depends on the presence of the M protein in the strain. Strains which do not possess the M protein at the time of typing cannot be typed by this method. In current laboratory practice a large proportion of strains may lack this antigen. It is recommended that freshly isolated strains be typed; passage and subculture in artificial media reduce the production of the M protein.

## D. T typing

The T typing system according to Griffith (1934) based on the agglutination reaction, is used to identify the strain according to the T protein. Since, however, various streptococcus types have more T antigens in common, the T typing result is only occasionally type-specific and so is merely referred to as a certain "T pattern".

Several technical requirements must be strictly observed in order to ensure reliable T typing. They are: the preparation of a sufficiently smooth suspension of the strain, sufficiently prolonged enzyme digestion to avoid spontaneous or cross agglutination and the use of potent and properly absorbed sera. Rabbit anti-T sera are commercially available.

### 1. *Preparation of the typing suspension*

Inoculate the strain which is to be typed into 5 ml Todd-Hewitt broth and incubate at 30 °C overnight. Thirty °C is preferable, but in most cases incubation at 37 °C is also possible. If the strain is not sufficiently, grown prolong the incubation for another 24 h. Centrifuge and suck off the supernatant, leaving about 0·2 ml of the fluid in which to resuspend the bacteria.

First digestion: add two drops of 5% trypsin solution in phosphate buffer pH 7·8 or two drops of pancreatic extract. (The preparation of pancreatic extract according to Cole and Onslow: suspend 100 g pig pancreas — fat free and minced — in 300 ml of distilled water. Add 100 ml of absolute ethyl alcohol, stir well and keep at room temperature for 3 days, occasionally shaking the mixture. Filter at first through muslin and afterwards through filter paper. Add 0·4 ml of concentrated HCl. If precipitate has developed filter through paper. Store the clear extract at +4°C.) Add one drop of Phenol red solution (composition as in Section III, C.1(a)) and adjust pH to 8·0, using first N NaOH, later for final adjustment N/5 NaOH. Place in a water bath for 1 h at 37°C. If the suspension is smooth, carry out the agglutination at pH 7. If the suspension after the first trypsinisation is granular, or if it is homogenous but is agglutinated nonspecifically by several sera, it must be subjected to a second or even third digestion by trypsin or pancreatic extract.

Second trypsinisation: Add one drop of pancreatic extract or trypsin solution and continue the digestion at pH 8 in a water bath at 37°C for 1 h. Bring the pH to 7·0 with N/5 HCl.

Third trypsinisation: Add one drop of pancreatic extract or trypsin solution and digest the suspension at pH 8 in a water bath at 50°C for 20 min. Bring the pH to 7·0 with N/5 HCl.

If the suspension after the third trypsinisation gives non-specific agglutinations, the preparation of the culture must be repeated; in such a case it is convenient to inoculate several cultures selected from different colonies. If agglutinability is lost after the digestion, a fresh culture must also be prepared and the procedure repeated, preferably with a shorter incubation with the pancreatic extract or trypsin solution.

## 2. *Performance of the agglutination reaction*

The agglutination reaction is carried out on a slide; using a platinum loop (diameter 2–3 mm) apply one drop of strain suspension and one drop of serum and mix. First use the polyvalent and then the corresponding monovalent sera.

The set consists of the following sera:

| Polyvalent | T serum Monovalent | | | |
|---|---|---|---|---|
| T | 1 | 3 | 13 | $B_{3264}$ |
| U | 2 | 4 | 6 | 28 |
| W | 5 | 11 | 12 | 27  44 |
| X | 8 | 14 | 25 | Imp 19 |
| Y | 9 | 22 | 23 | |

### 3. *Evaluation of the results*

The reaction is evaluated within 1 min, or at maximum within 2 min; immediate agglutination is marked + + +, strong reaction within 1 min + + and weak but distinct agglutination within 1–2 min +.

Agglutination reactions may develop in one strain with several sera, but these sera must be of those types which belong to one T pattern.

The T patterns described so far are as follows:

| T patterns | M types implied |
|---|---|
| (a) corresponding to several M types: | |
| 3/13/B$_{3264}$ | 3, 13, 33, 39, 41, 43, 52, 53, 56 |
| 4/28 | 4, 24, 26, 28, 29, 46, 48, 60 |
| 5/12/27 | 5, 11, 12, 27, 44, 61 |
| 14/41 | 14, 49 |
| 15/23/47 | 15, 17, 19, 23, 30, 47, 54 |
| 8/25/Imp 19 | 2, 8, 25, 31, 55, 57, 59 |
| (b) corresponding to one M type: | |
| 1 | 1 |
| 6 | 6 |
| 9 | 9 |
| 18 | 18 |
| 22 | 22 |

At present no adequate information is available on the T patterns of M types which are not included in this table.

It is advisable to use the T typing procedure in classification of group A strains. The T sera are commercially available and the information on the T pattern of the strain is useful in epidemiological work and quite frequently also in diagnostic practice.

## V. TYPING OF GROUP B STREPTOCOCCI

### A. Importance of type identification

In recent years the great interest devoted to group B streptococcus has resulted in evidence that this pathogen may play an important role in human infections. Various clinical patterns of infections due to group B streptococcus have been identified, the most important being those of the urogenital tract of women and infections of newborns (neonatal septicaemia and meningitis) (Jelínková, 1963; Wilkinson *et al.*, 1973).

These findings have considerably stimulated studies of group B strepto-

coccus antigenic structure. The procedures for typing group B streptococci have been refined (Lancefield, 1934; Wilkinson and Eagon, 1971;
Wilkinson, 1975). Type identification is essential for the epidemiological
tracing of group B infections. Furthermore, there are indications that the
type antigens are the factors of virulence of the group B streptococcus and
that the antibodies to them provide immunity.

The differentiation of five types within group B streptococcus is based on
the scheme of Lancefield (1934). The antigenic structure of these types is
as follows (Wilkinson, 1975).

| Type | Antigenic structure | |
|------|---------------------|-----------------|
|      | Polysaccharide antigen | Protein antigen |
| Ia   | Ia                  |                 |
| Ib   | Ib                  | Ibc†            |
| Ic   | Ia                  | Ibc             |
| II   | II                  | Ibc (occasionally) |
| III  | III                 | Ibc (rarely)    |

† Formerly Ic.

Some strains of type III carry the R protein antigen. This antigen, as
well as another protein antigen designated X, form, for some authors
(Pattison *et al.*, 1955), the basis of distinguishing types R and X in group
B streptococci in addition to the types Ia to III.

## B. Serological typing by precipitation reaction

The type identification of group B streptococcus is performed by a
precipitation reaction in a capillary tube or as a ring test. Absorbed type-
specific hyperimmune rabbit serum and an extract of the type antigen
from the cell of the streptococcus to be typed are used (Jelínková, 1963).
The typing sera are not commercially available. They are produced by a
few advanced laboratories only. However, they can be produced in any
laboratory having adequate knowledge of group B streptococcus bacteriology.

### 1. *Preparation of typing extract*

The procedure is the same as indicated in Section III. C, 1 (b) for group
A streptococci (Lancefield extract), except that the extraction with N/5
HCl is performed at 50–52 °C for 2 h.

### 2. *Performance of the reaction*

Two procedures may be used:

(a)   a ring test in conically narrowing capillaries
(b)   precipitation in capillary tubes.

Both procedures are the same as described for streptococcus grouping in Section III.C, 2.

### 3. Evaluation of results

A prerequisite for obtaining reliable results is the use of properly and adequately absorbed typing sera. Clear-cut reactions should be recorded only. It is advisable to use all five typing sera for typing each strain.

## REFERENCES

Becker, C. G. (1967). *Proc. Soc. exp. Biol.*, **124**, 331–335.
Caravano, R. (Ed.) (1968). *Current Research on Group A Streptococcus.* A Symposium. Excerpta Medica Foundation, Paris.
Davies, H. C., Karush, F., and Rudd, J. H. (1968), *J. Bact.*, **95**, 162–168.
Facklam, R. R. and Moody, M. D. (1968). *Appl. Microbiol.*, **16**, 1822–1825.
Fox and Wittner (1968).
Fuller, A. T. (1938). *Br. J. exp. Path.*, **19**, 130–139.
Ginsburg, I., and Grossowicz, N. (1957). *Proc. Soc. exp. Biol. Med.*, **96**, 108–112.
Goldin, M., and Glenn, A. (1964). *J. Bact.*, **87**, 227–228.
Griffith, F. (1934). *J. Hyg.*, **34**, 542–584.
Haverkorn, M. (Ed.) (1973). *Streptococcal Disease and the Community.* A Symposium. Amsterdam.
Jelínková, J. (1963). *Cesk. Epidemiol. Mikrobiol. Imunol.*, **12**, 74–80.
Kaplan, M. H. (1963). *J. Immunol.*, **90**, 595–606.
Kodíček, S., and Mistry, S. P. (1955). *Arch. Bioch. Biophys.*, **4**, 30–36.
Kunter, C. (1963). *Zbl. Bakt.*, **188**, 190–194.
Lancaster, L. J., and Sherris, J. C. (1960). *Am. J. Clin.*, **34**, 131–132.
Lancefield, R. C. (1928). *J. exp. Med.*, **47**, 91–103.
Lancefield, R. C. (1933). *J. exp. Med.*, **57**, 571–595.
Lancefield, R. C. (1934). *J. exp. Med.*, **59**, 441–458.
Lancefield, R. C. (1943). *J. exp. Med.*, **78**, 465–476.
Lancefield, R. C. (1957). *J. exp. Med.*, **106**, 525–544.
Lancefield, R. C. (1958). *J. exp. Med.*, **108**, 329–341.
Lancefield, R. C., McCarty, M. and Everly, W. W. (1975). *J. exp. Med.*, **142**, 165–179.
Lyampert, I. M., Vedenskaja, O. V., and Danilova, T. A. (1966). *Immunology*, **11**, 313–320.
Maxted, W. R. (1948). *Lancet*, **2**, 255–256.
Maxted, W. R. (1949). *J. Gen. Microb.*, **3**, 1–6.
Maxted, W. R. (1953). *J. Clin. Path.*, **6**, 224–226.
McCarty, M. (1959). *J. exp. Med.*, **109**, 361–378.
McLean, S. J. (1953). *J. gen. Microb.*, **9**, 110–118.
Michael, J. B., and Massel, B. F. (1965). *J. Lab. Clin. Med.*, **65**, 322–328.
Moody, M. D., Ellis, E. C., and Updyke, E. L. (1958). *J. Bact.*, **75**, 553–560.
Moody, M. D., Padula, J., Lizana, D., and Hall, C. T. (1965). *Hlth Lab. Sci.*, **2**, 149.

Munch-Petersen, E., and Christie, R. (1947). *J. Path. Bact.*, **59**, 367–371.
Pattison, I. H., Matthews, P. R. J., and Maxted, W. P. (1955). *J. Path. Bact.*, **69**, 43–50.
Petit, J. F., Munoz, E., and Ghuysen, J. M. (1966). *J. Biochem.*, **5**, 2764.
Rantz, L. A., and Randall, E. (1955). *Stanford Med. Bull.* **13**, 290–291.
Redys, J. J., Parzick, A. B., and Borman, E. K. (1963). *Publ. Hlth Rep.*, **78**, 222–226.
Rotta, J. (1970). *Group Identification of Haemolytic Streptococci.* A laboratory pamphlet.
Rotta, J. (1975). *Type Identification of Haemolytic Streptococci.* A laboratory pamphlet.
Rotta, J., Krause, R. M., Lancefield, R. C., Everly, W., and Lackland, H. (1971). *J. exp. Med.*, **134**, 1298–1315.
Stewart, W. A. *et al.* (1944). *J. exp. Med.*, **79**, 99–114.
Swift, H. F., Wilson, A. T., and Lancefield, R. C. (1943). *J. exp. Med.*, **78**, 127–134.
Swanson, J., Hsu, K. C., and Gotschlich, E. C. (1969). *J. exp. Med.*, **130**, 1063–1091.
Wagner, M., and Heinrich, H. (1962). *Zbl. Bakt.*, **186**, 292–302.
Wallerström, A. (1962). *Acta path. microb. Scand.*, **56**, 459–464.
Widdowson, J. P., Maxted, W. R., and Grant, D. L. (1970). *J. gen. Microb.*, **61**, 343–353.
Widdowson, J. P., Maxted, W. R., and Pinney, A. M. (1971). *J. Hyg. Camb.*, **69**, 553–564.
Wilkinson, H. W., and Eagon, R. G. (1971). *Infect. Immun.*, **4**, 596–604.
Wilkinson, H. W., Facklam, R. R., and Wortham, E. G. (1973). *Infect. Immun.*, **8**, 228–235.
Wilkinson, H. W. (1975). *Inf. Immun.*, **11**, 845–852.
Willey, G. G., and Bruno, P. N. (1968). *J. exp. Med.*, **128**, 959–968.
Wolley, D. W. (1941). *J. exp. Med.*, **73**, 487–493.

# CHAPTER IX

# Identification and Typing of Enterococci

J. Jelínková

*Postgraduate Medical and Pharmaceutical Institute,*
*Department of Medical Microbiology, Prague*

AND J. Rotta

*Institute of Hygiene and Epidemiology, Prague*

## I. INTRODUCTION

To taxonomically place the Gram-positive streptococci, present in the faeces of humans and warm-blooded animals in the early days of medical microbiology (Thiercelin, 1899) established the taxon "enterococci". The enterococci were characterised by morphological features, several cultural properties and source of isolation. When information was later provided on the antigenic structure, biochemical and physiological properties, and on the pathogenic role of the "enterococci" (as well as "non-enterococci" and "enterococcus-like" bacteria) for man and animals, further designations were introduced, viz. faecal streptococci and group D streptococci. These terms are often erroneously used as synonyms for enterococci (Hartman *et al.*, 1966).

With the knowledge presently available it is possible to devise a useful classification scheme which enables identification of most important pathogens in this group of bacteria. Although the scheme is not ideal from the strictly taxonomic point of view, it is quite helpful in routine practice. It should, however, be pointed out, in this connection that "enterococci" differ in a number of morphological, antigenic and physiological characters from the rest of the cocci belonging to the genus *Streptococcus*. This fact has been a strong argument in favour of establishing a new genus for the enterococci.

## II. DEFINITION AND CLASSIFICATION

### A. Definition of enterococci

Micro-organisms classified as enterococci are spherical, slightly elongated along the axis of adherent cells. Cell diameter is $0\cdot5$–$1\cdot0\,\mu$m. The cells are arranged in pairs or short chains. Enterococci grow rapidly in the temperature range from 10 °C up to over 40 °C, some species even growing at 50 °C. They are heat tolerant to 60 °C for 30 min. They grow in media containing $6\cdot5\%$ NaCl, $40\%$ bile, at pH $9\cdot6$, and reduce litmus in milk (Sherman, 1937, 1938). They tolerate the presence of some substances in concentrations inhibitory for other cocci. Enterococci metabolise a large number of substrates, and may be motile or non-motile. Enterococci belong to Lancefield's group D streptococci, which contains the glycerol type of teichoic acid, the group-D-specific antigenic determinant. Enterococci are widely distributed in nature and easily colonise human beings and animals. Although they are usually saprophytic, they may produce acute or mild localised processes or serious systemic infections (Duma *et al.*, 1969; Hoppes and Lerner, 1974; Watanakunakorn, 1974; Parker and Ball, 1976).

### B. Classification

The classification of enterococci is based on the source of isolation, and morphological, physiological, and biochemical characters. At present, the following classification scheme is used (Fig. 1). The notes and comments which follow immediately below relate to some of the so-far unsettled problems of classification and interspecies relations:

(a) *S. faecalis* is at present divided into three varieties based on haemolytic and proteolytic characters only. Drucker and Melville (1973) by numerical taxonomy could not verify the differentiation into three varieties. Other similar studies have only found two distinct enterococcal clusters: *S. faecalis* and *S. faecium* (Jones *et al.*, 1972).

(b) Bergey's Manual of Determinative Bacteriology (8th ed, 1974) classifies *S. durans* under *S. faecium* and does not list it as a separate species. *S. faecium* has priority as a single species name (Colman, 1970).

(c) In recent years, *S. bovis I* and *S. bovis II* have been differentiated (Maxted and Parker, 1973, personal communication; Facklam, 1973; Parker and Ball, 1976 — see section IV.B.6). *S. bovis* I is often associated with endocarditis.

(d) *S. suis* (streptococcus group S, De Moor, 1963) is related to *S. bovis* (Elliott, 1966).

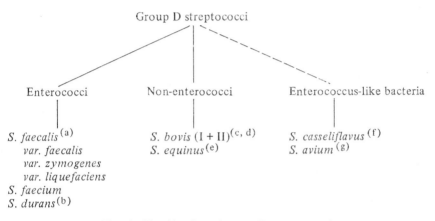

FIG. 1. Classification of group D streptococci.

(e) *S. equinus* has not been isolated from humans.

(f) *S. casseliflavus* (Mundt and Graham, 1968) is most often considered a variety of *S. faecium*. It is an epiphyte on plants, gives a very weak group D reaction, survives 60 °C for 30 min, but does not grow at 45 °C, and produces a water-soluble, pale yellow pigment. In some physiological characteristics it resembles *S. faecalis*.

(g) *S. avium* (streptococcus group Q, Guthof, 1955; Nowlan and Deibel, 1967) is physiologically closely related to enterococci. Group Q and D antigens may be present in the same strain (Smith and Shattock, 1964). *S. avium* has not been isolated from humans.

## III. BASIC BIOLOGICAL CHARACTERISTICS OF ENTERO-COCCI

### A. Physiological characteristics and growth requirements

Enterococci utilise a large number of polysaccharides as energy source. For example, most *S. faecalis* and *S. faecium* strains produce acid from glucose, fructose, lactose, sucrose, maltose, cellobiose, trehalose, galactose, and particular species metabolise some further substances, mannitol, sorbitol, glycerol etc. Cultivation of enterococci in glucose broth results in a final pH between 4·0 and 4·6. Enterococci are catalase negative. However, a catalase-like activity has been found in some strains (Jones *et al.*, 1964).

Enterococci require 7–14 amino-acids and 5–6 B vitamins. Although the presence of purines and pyrimidines is not essential, their addition to growth media enhances growth (Deibel *et al.*, 1963).

The yellow pigmentation of some enterococci depends on the production of carotenoids (Taylor *et al.*, 1971). They have been identified in *S. faecium* as a new class of terpenoids, namely the triterpenoid carotenoids (Taylor and Davies, 1974a, b).

Enterococci tolerate the presence of any of the following substances: 6·5% NaCl, 40% bile, 0·1% thallous acetate, 0·02% sodium azide, 0·5–1·0 units of Penicillin per 1 ml of culture medium; they grow at pH 9·6. Strains of *S. faecalis* grow in the presence of 0·04% tellurite, while *S. faecium* strains are inhibited.

Enterococci may produce bacteriocins which in exceptional instances act on *Streptococcus pneumoniae*, *Clostridium perfringens*, *Streptococcus salivarius* and *Listeria monocytogenes* (Bottone *et al.*, 1974; Krämer and Brandis, 1975). Bacteriocin-producing strains of *S. faecalis* more often produce haematogenous pyelonephritis in the rat (Montgomerie *et al.*, 1973).

Failure to neutralise the activity of bacteriocins produced by *S. faecium* with antiphage sera suggests that there is no relationship between temperate phages and bacteriocins (Krämer and Lenz, 1975). Identity of a bacteriocin with a haemolysin in *S. faecalis var. zymogenes* strains is presumed (Brock and Davie, 1963); both characters are plasmid-determined (Jacob *et al.*, 1975).

The occurrence of motile enterococci is rather rare, but such strains may be isolated from human as well as animal material. Motile strains possess one or two, exceptionally three or four flagella (Jelínková and Rýc, 1971). Motile streptococci display a chemotactic affinity towards sugars and amino-acids (Van der Drift *et al.*, 1975).

An intracellular structure of unknown physiological role, but of distinct specificity for group D streptococci, has been described in enterococci (McCandless *et al.*, 1968, 1972). The structure is a long, thin, cylinder-like core located in the cytoplasm.

L forms of enterococci are easily isolated (Gooder, 1968; Young and Armstrong, 1969; King and Gooder, 1970).

Cultivation of enterococci on solid and in liquid media is easy. Numerous selective media for presumptive differentiation or identification of entero-cocci in routine practice have been described (Facklam, 1973, personal communication; Efthymiou and Joseph, 1974; Daoust and Litsky, 1975). Further information and descriptions of laboratory procedures are given in Section IV.

## B. Antigenic structure

The group D-specific antigen in enterococci as well as in other bacterial species belonging to this group is a teichoic acid which is a polymer of glycerol phosphate having some of the hydroxyl groups of glycerol substituted by D-alanyl and kojibiosyl (2-O-$\alpha$-D-glucopyranosyl-(1$\rightarrow$2)-D-glucopyranosyl) residues (Elliott, 1959; Toon *et al.*, 1972). Attempts to locate the group-specific antigen in the cell suggest that it is situated in the cytoplasm (Elliott, 1959), or between the cytoplasmic membrane and the cell wall (Hay *et al.*, 1963; Smith and Shattock, 1964). Recent studies (Bauer *et al.*, 1974; Garland *et al.*, 1975) using electron microscopy have shown that teichoic acid is most likely not intimately associated with the cell wall. However, there is evidence that teichoic acid is associated with cell wall, and cytoplasmic membrane.

Teichoic acid isolated from *S. faecalis* precipitates with type XII antipneumococcal sera; this cross-reactivity is due to the presence of kojibiosyl residues in both antigens. Another immunological relationship of enterococcus teichoic acid was established with type XVI pneumococcus and probably depends on the presence in both antigens of poly-glycerol phosphate and D-glucose residues (Heidelberger and Baddiley, 1974).

The type-specific antigens in enterococci as well as in other group D streptococci are polysaccharide by nature. They are situated in the cell wall. These type-specific polysaccharides are thus counterparts of the group-specific antigens in other Lancefield groups of beta-haemolytic streptococci.

The polysaccharides of type 1 and 26 contain rhamnose, glucose, galactose, N-acetylglucosamine and N-acetylgalactosamine (Bleiweis and Krause, 1965). In type 1 polysaccharide, the D-glucose and N-acetyl-glucosamine are most likely components of the antigenic determinant.

Another serologically active component associated with the type-specific antigens in group D streptococci has been described (Krause, 1972).

This is a rhamnose polymer that gives strong cross-reactions with group A-variant cell-wall polysaccharide (Elliott *et al.*, 1971).

The main structural component of the cell wall in enterococci is peptidoglycan. Although no information has been obtained about its immunological behaviour so far, the peptidoglycan of *S. faecalis*, *S. liquefaciens* (Slade and Slamp, 1972) and *S. zymogenes* is chemically different from *S. faecium* and *S. durans*. *Streptococcus bovis* and *S. equinus* differ from enterococci by the presence of threonine in their peptidoglycan.

## IV. IDENTIFICATION METHODS

### A. Cultivation of specimens

Most of the solid and liquid media in current use allow reliable cultivation of enterococci from specimens by overnight incubation at 37 °C. In examining human materials such as stools, urine, throat swabs, blood, sputa, etc., the use of *blood agar* allows ready evaluation of colony morphology and haemolysis type.

### B. Strain identification

The differentiation of isolates conventionally should procede stepwise in the following order (Jelínková and Mottl, 1975):
Differentiation of group D streptococci from other groups.
Differentiation of group D streptococci into enterococcus and non-enterococcus species (Fig. 2).
Differentiation into individual species and varieties (Table I).

### 1. *Differentiation of group D from other groups of streptococci*

This implies serological demonstration of the group-specific antigen. However, some group D strains show weak precipitation reactions with usual diagnostic sera. The reason may lie in the quality of the serum batch employed, in weak production of the group antigen in the strain (e.g. *S. bovis*, *S. equinus*), or in the location of the D antigen in the cell. In these cases (and also in laboratories which do not group streptococci serologically), the "bile-esculin" screening test which is only positive in group D strains, may be performed as a substitute. An auxiliary criterion for group D strains is growth in broth at 45 °C.

### (a) *Serological grouping*

*Preparation of group antigen: Lancefield's method* (1933): add 0·5 ml of N/5 HCl to bacteria obtained by centrifugation of 80 ml of a 16 h culture in glucose broth without phosphates, pH 7·0–7·1 (meat peptone broth

## TABLE I

### Differentiation of enterococci

| | 6·5 % NaCl | Beta haemolysis | Gelatin | Arabinose | Mannitol | Sorbitol | Arginine | Sucrose | Inulin | Lactose |
|---|---|---|---|---|---|---|---|---|---|---|
| *S. faecalis var. faecalis* | + | − | − | − | + | + | + | + | − | + |
| *var. liquefaciens* | + | − | + | − | + | + | + | + | − | + |
| *var. zymogenes* | + | + | − (+) | − | + | + | + | + | − | + |
| *S. faecium* | + | − | − | + | + | − | + | + | − | + |
| *S. durans* | + | − (+) | − | − | − | − | + | − (+) | − | + |
| *Non-enterococci* | | | | | | | | | | |
| *S. bovis* (I + II) | − | − | − | − (+) | ±† | − | − | + | ±† | + |
| *S. equinus* | − | − | − | − (+) | − | − | − | + | − | − |

† In *S. bovis* II usually negative.
*S. bovis* I dextran production +, starch hydrolysis +.
*S. bovis* II dextran production −, starch hydrolysis −.

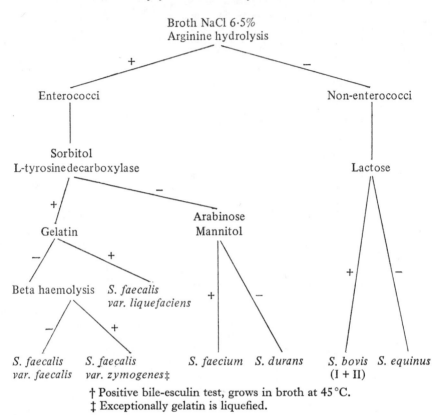

† Positive bile-esculin test, grows in broth at 45 °C.
‡ Exceptionally gelatin is liquefied.

FIG. 2. Identification scheme for group D streptococcus species.†

with 1% glucose, Formula 1), shake well and place in a boiling water bath for 10 min. Cool, add one drop of Phenol red (Formula 2) as indicator, and neutralise with 1 N NaOH; centrifuge; the clear supernatant contains the group-specific antigen.

*Fuller's method* (1938): add 0·1 ml of formamide (HCONH₂) to bacteria obtained by centrifugation of 10 ml of a 16-h culture in Todd-Hewitt broth (Difco, Oxoid). Shake well and place in an oil bath at 150 °C for 10 min. Cool, add 0·25 ml of acid alcohol (1 ml of concentrated HCl in 99 ml of C₂H₅OH 95%), shake, centrifuge and add 0·5 ml of acetone to the clear supernatant. A precipitate of the group-specific antigen will form, which is separated by centrifugation and dissolved in 0·35 ml of saline. Add a drop of Phenol red and neutralise with N/5 NaOH.

*Precipitation reaction:* the precipitation reaction is performed in capillaries, conically narrowed and sealed at one end. The diameter of the wider part

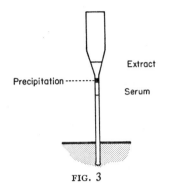

FIG. 3

is about 0·5 cm, total length 3–4 cm (the shape is shown in Fig. 3). The capillary is fixed in a groove filled with plasticine in a rack. Use a Pasteur pipette first to deliver serum up to the middle of the narrower part of the capillary. Then add extract on top, taking care that the two liquids do not mix. If the reaction is positive, a precipitation ring forms on the interface between the two reactants. The reading is performed against a dark background within 5 min.

At the same time carry out a positive control with an extract from reference *S. faecalis* strain and a negative control with serum of another group.

Glucose broth (Formula 1)
　　Add 10 g of glucose (1%) to 1000 ml of meat-peptone broth (pH 7·0–7·1), dissolve hot, dispense into flasks, sterilise at 100°C in steam for 3 × 20 min.

Phenol red solution (Formula 2)
　　0·1 N Solution of sodium hydroxide NaOH 25 ml
　　　Phenol red　　　　　　　　　　　　1 g
　　　Distilled water　　　　　　　　　475 ml

Place the solution into an incubator at 37°C for 1–2 days until the dye is completely dissolved, shaking thoroughly several times a day. Then filter through filter paper. Do not sterilise!

*A fluorescent antibody direct staining method* for serological grouping has also been described for group D streptococci (Pavlova *et al.*, 1972), but is not widely used because suitable commercial diagnostic products are not yet available.

*Counter-immunoelectrophoresis* has been introduced into serological identi-

8

fication of group D enterococcus strains (Portas *et al.*, 1976), but in group D non-enterococcus strains CIE has not given satisfactory results.

(b) *Bile-esculin screening test.* The test makes use of, jointly, tolerance to 40% bile and esculin hydrolysis (esculitin being detected by ferric citrate).

Inoculate the culture in a wavy line across a slanting agar surface and by stabbing. Read after 24–48 h of incubation. A positive test is indicated by growth and blackening.

Bile-esculin medium (Formula 3)
 Nutrient agar 40† g in 400 ml of water.
 Ox bile 400 ml (or equivalent concentration of Difco Dehydrated ox gall).
 Ferric citrate ($Fe^{3+}$) 0·5 g in 100 ml of water.
 Esculin 1 g in 100 ml of water.

Dissolve the nutrient agar in 400 ml of water by heating at 100 °C in running steam, add the bile and citrate solution and sterilise for 30 min by heating at 121 °C in steam under pressure. After cooling to about 50 °C, add, under sterile conditions, 100 ml of 1% esculin solution in distilled water, sterilised by filtration through a bacterial filter. After stirring thoroughly dispense the medium (7–10 ml) into test-tubes under sterile conditions and allow to solidify in a slant position.

(c) *Test for growth in broth at* 45 °C. Place the broth into a 45 °C water bath immediately after inoculation. Evaluate growth daily for 1–5 days.

The test is positive in all group D strains, but also in strains of some other streptococcal species (groups Q, E, *S. milleri, S. thermophilus*). Thus a negative result precludes the strain from being an enterococcus or non-enterococcus.

## 2. *Differentiation of group D strains into enterococcus and non-enterococcus species*

Enterococci: positive growth in broth containing 6·5% NaCl, positive hydrolysis of arginine.
Non-enterococci: negative growth in broth containing 6·5% NaCl, negative hydrolysis of arginine.

(a) *Growth in* 6·5% *NaCl broth.* Use only fresh medium and perform a parallel control with a reference strain of *S. faecalis.* Only obvious growth is evaluated; take readings daily for 1–5 days.

† Depends on the individual product.

6·5% NaCl broth (Formula 4)

To meat-peptone broth (containing 0·5% of NaCl), add sodium chloride to a final concentration of 6·5%. Allow the salt to dissolve, dispense the broth in 5 ml amounts into test-tubes and autoclave at 121 °C for a period of 20–30 min.

(b) *Hydrolysis of arginine.* Inoculate medium and overlay with sterile paraffin oil. Perform first reading after 18–24 h incubation, not sooner. Growth and violet colouring signify a positive result. Growth and yellow colouring signify a negative result.

Falkow (1958) medium for decarboxylase detection (Formula 5)

| Peptone | 5 g |
|---|---|
| Yeast extract | 3 g |
| Glucose | 1 g |
| Distilled water | 1000 ml |
| Bromcresol purple | 10 ml |

Dissolve ingredients in water, adjust pH to 6·7, then add indicator and 0·5 L-arginine. Check that pH equals 6·7 and distribute medium into test-tubes (1 ml). Sterilise at 115 °C for 20 min.

3. *Differentiation of enterococci into species*

S. faecalis (and varieties):  positive sorbitol
                              positive L-tyrosine decarboxylase
S. faecium and S. durans:  negative sorbitol
                           negative L-tyrosine decarboxylase

(a) *Production of L-tyrosine decarboxylase.* A modification of Mead's medium is recommended for detection of the enzyme (Mead, 1963). This is a solid selective diagnostic medium for the identification of S. faecalis and its varieties. The selectivity of the medium is due to the presence of thallium acetate: the identification principle is based on the ability of S. faecalis and its varieties to ferment sorbitol and produce L-tyrosine decarboxylase, which is active only in the presence of an acid; in the present case produced by sorbitol fermentation.

Inoculate the culture by a stab and on the surface of a segment of the medium.

Read after 18–24 h of incubation. Strains with positive decarboxylase activity produce a translucent zone, colonies are white.

Mead's medium (modified) (Formula 6)

| | |
|---|---|
| Peptone | 10 g |
| Yeast extract | 1 g |
| Sorbitol | 2 g |
| L-tyrosine | 5 g |
| Thallium acetate | 0·05 g |
| Agar | 20 g |
| Distilled water | 1000 ml |

Dissolve the ingredients one by one (except thallium acetate and agar) in hot water and adjust pH to 6·2. Add agar and heat in running steam until it has dissolved. Sterilise in autoclave at 115 °C for 10 min.

Add thallium acetate (50 mg per 1000 ml of medium) and check pH (6·2). Decant 300 ml from 1000 ml of this medium and add another 4·5 g of L-tyrosine to the decanted portion.

Distribute the remaining 700 ml of medium (without raising concentration of L-tyrosine) into small Petri dishes. After the medium has solidified, overlay the plates with a second thin layer of medium, that with the raised L-tyrosine concentration, cooled almost to the point of solidification to ensure that undissolved tyrosine remains evenly dispersed.


(b) *Sorbitol fermentation*. For all remaining fermentation activity tests, the medium described below (Formula 7) is used. Hiss's serum water (Cowan and Steel, 1974) may also be used as a medium basis.

Evaluate obvious growth and change of indicator. Perform readings daily for a period of 1–5 days. After recording the results, check all tubes by subculture on blood agar.

Fermentation activity medium (Formula 7)

| | |
|---|---|
| Peptone | 10 g |
| NaCl | 5 g |
| $K_2HPO_4$ | 1 g |
| Distilled water | 1000 ml |

Dissolve the ingredients in hot water and adjust pH to 7·2–7·4. Boil for a short time in running steam (100 °C) and, if necessary, filter through paper. To 100 ml of this base add 1 g (1%) of the particular carbohydrate (inulin and esculin only 0·5%) and 1·5 ml of Bromothymol blue solution as indicator of pH change. Then dispense the medium into test-tubes and sterilise at 100 °C in running steam for 15 min on three consecutive days.

Preparation of indicator:
0·1 N Solution of sodium hydroxide, NaOH  25 ml
Bromthymol blue 1 g
Distilled water 475 ml

Place the solution in the incubator at 37°C for 1–2 days until the dye has dissolved completely. During this period shake the solution thoroughly several times a day. Filter through paper. Do not sterilise.

### 4. *Differentiation of the species* S. faecalis *into varieties*

*var. faecalis:* not beta-haemolytic, gelatinase not produced.
*var. zymogenes:* beta-haemolytic, gelatinase mostly not produced.
*var. liquefaciens:* not beta-haemolytic gelatinase produced.

(a) *Type of haemolysis:* is inspected on blood agar plate after overnight incubation.

(b) *Gelatin liquefaction.* Smith and Goodner's (1958) medium, originally introduced for the diagnosis of vibrios is recommended.
    Inoculate culture abundantly by stabbing, read first after 20 h incubation, and finally after storage at 4°C for 18 h (the reaction becomes more obvious)
    Evaluate the zone of milky translucence around the stab. In parallel, set up a positive control with a reference strain of *S. faecalis var. liquefaciens.*

Medium for gelatinase detection (Formula 8)
    Peptone (neopeptone or proteose type)     4 g
    Yeast extract                             1 g
    Agar                                      2 g
    Distilled water                        1000 g

Dissolve the ingredients by heating at 100°C in running steam. Adjust pH to 7·3–7·4 and add 30 g (3%) of gelatin. Heat for a short time in running steam for the gelatin to dissolve, sterilise for 15 min by heating at 121°C in steam under pressure. Distribute the medium into Petri dishes.

### 5. *Differentiation between* S. faecium *and* S. durans

*S. faecium:* positive arabinose, and mannitol
*S. durans:* negative arabinose, and mannitol

(a) *Fermentation activity tests:* see Section IV.B, 3. (b). Use medium Formula 7. Table I.

### 6. Differentiation between non-enterococcal species

*S. bovis* I: positive lactose, mannitol, inulin, starch hydrolysed, dextran produced from sucrose

*S. bovis* II: positive lactose, negative mannitol (mostly) and inulin (mostly), starch not hydrolysed, dextran not produced

*S. equinus:* negative lactose

(a) *Fermentation activity tests:* see Section IV.B, 3 (b). Use medium Formula 7.

(b) *Starch hydrolysis.* The procedure recommended by Cowan and Steel (1974) is used. Inoculate starch agar plate (Formula 9) and incubate at 37 °C for 2–4 days.

Flood the agar plate with Lugol's iodine solution (Formula 10). The medium turns blue where starch has not been hydrolysed. Hydrolysis is indicated by clear colourless zone. Positive control is necessary.

Starch agar (Formula 9)

| | |
|---|---|
| Potato starch | 10 g |
| Distilled water | 50 ml |
| Nutrient agar 2% | 1000 ml |

Triturate the starch with the distilled water to a smooth cream and add to the molten nutrient agar. Mix, and sterilise at 115 °C for 10 min. Distribute into small Petri dishes. The medium should not be filtered after the starch suspension has been added. Overheating may hydrolyse the starch.

Lugol's iodine solution (Formula 10)

| | |
|---|---|
| Iodine | 5 g |
| Potassium iodide | 10 g |
| Distilled water | 100 ml |

For use dilute 1:5 with distilled water.

(c) The method used for the production of extracellular dextran from sucrose, recommended by Parker and Ball (1976), is a modification of Hehre and Neill's (1946) based on the differential precipitation of dextran and levan (produced in sucrose broth) with ethanol.

## V. PRINCIPLES OF ENTEROCOCCUS TYPING

In searching for sources of infection or contamination of foods or water, a more detailed differentiation between struck is often needed. This demand is partly met by type classification based on serological demon-stration of type-specific antigens or typing with the use of bacteriophages or bacteriocins. These approaches to typing bear no direct relationship to individual enterococcus species and varieties.

## A. Serological typing

### 1. Introduction

Work undertaken to elaborate a system of serological enterococcus typing so far has been rather limited. A practicable typing method for *S. faecalis* was elaborated by Sharpe and Shattock (1952). Maxted and Fraser (1964) modified it in order to improve the typability of *S. faecalis*, *S. zymogenes*, and *S. liquefaciens* and enhance the reproducibility of results. More than 30 types are known, but their taxonomic position may be subject to revision depending on new immunochemical findings. For example, in England (Maxted and Fraser, 1964) 94% of *S. faecalis* strains from urine specimens have been successfully typed, the predominating serotypes being 1, 9 and 19. In Roumania (Pleceas, 1967; Pleceas *et al.*, 1968), 96% of strains isolated from humans, animals, food or water could be serotyped. In *S. faecalis* mostly the *S. zymogenes* variant, isolated from urinary infections serotypes 1, 9/19, and 9/19/41 have dominated.

### 2. Type-specific polysaccharide

Type-specific polysaccharide is extracted from cells of the strain to be typed by the following procedure: bacteria from 50 ml of an over-night culture grown at 37°C in nutrient broth containing 0·5% of glucose are harvested by centrifugation. Add 0·5 ml of N/20 HCl to the deposit of bacteria and heat for 10 min in a boiling water bath. Neutralise with N/5 NaOH to pH 7·0 using Phenol red solution as an indicator, centrifuge and collect the supernatant fluid which contains the type antigen (and the group antigen) ready for use in serological typing. If the antigen solution does not react with the typing sera, modify it as follows: to one volume of extract add three volumes of acid alcohol (99 ml of abso-lute ethyl alcohol and 1 ml of concentrated HCl), collect the precipitate and dissolve in a half or a quarter volume of the original extract. This procedure both concentrates the extract and separates the alcohol soluble inhibitory substance interfering in the precipitation.

### 3. Antisera

Antisera used for serological typing should be hyperimmune rabbit sera obtained by 4 to 8-week immunisation of rabbits with formalin killed vaccines. The sera must be rendered type-specific by absorption with bacteria of a heterologous type (bacteria killed at 80°C for 30 min). One or more absorptions are used employing a ratio of 1 volume of packed cells and three volumes of serum, the mixtures being incubated for 30 min at 37°C.

### 4. Serological typing

Serological typing is best performed as a ring test or a capillary test. In the former test, the reading is carried out within 5 min at room temperature, in the latter test after 30 min of incubation at 37°C.

## B. Phage typing

### 1. General principles, lysogeny and phage sensitivity

Bacteriophages were first observed in enterococci by Beckerich and Hauduroy (1922). They were not active against other streptococcal species. Their specificity for enterococci was described by Evans (1934), Graham and Bartley (1939), Evans and Chinn (1947) and by Kjems (1955).

Ciuca et al. (1959) proposed making use of the species specificity of enterococcal phage typing. They were the first to elaborate a phage-typing method for the differentiation of enterococci. In their experiments, three specific bacteriophages were employed. Two were active (separately or by combined action) against 92% of strains belonging to S. faecalis and its varieties, whereas the remaining phage worked against 35% of strains of S. faecium, and S. durans, or S. bovis (65% of strains tested were resistant). Phage-sensitivity testing was then recommended for rapid differentiation of enterococcus species.

A more sophisticated approach to the identification of individual enterococcal species by using mixtures of selected phages was introduced by Pleceas and Brandis (1974, 1975). These authors used four different pools of phages for identification: a pool of 23 phages for group D, a pool of 17 phages for S. faecalis and its varieties, a pool of three phages for S. faecium and S. durans and a pool of three phages for S. bovis. Out of over 400 strains tested, 90% were identified.

Phage typing has been employed, not only for the rapid identification of species, but also for better differentiation of individual strains of the same species. As in bacteriocin typing (see below), differentiation of strains by phage typing has mainly been performed by determining the sensitivity spectrum of the strain.

Determination of lysogeny has been used only exceptionally. Lysogeny has been demonstrated in a majority of strains. Very exceptionally, in a few strains only, production of phages active against the strains themselves has been found (Tzannetis et al., 1970). Repeated phage typing has shown that lytic patterns are reproducible (Hérmán and Hoch, 1971). On an average, 77% of group D strains are sensitive to phage typing (Pleceas *et al.*, 1974).

The frequency of phage-typable strains among enterococci differs between laboratories. This frequency depends on the set of phages used (Hoch and Hérmán, 1971), the origin of the phages and their stability upon storage. Membrane-stabilizing substances, human serum, magnesium and sucrose increase stability (Hérmán and Hoch, 1971). A classification of enterococci isolated from the human urogenital tract according to a new phage-typing scheme has been published by Caprioli *et al.* (1975). These authors used a set of seven phages for typing *S. faecalis* and *S. liquefaciens* into 27 types (90% were typable) and another set of ten phages for typing *S. faecium* and *S. durans* into 22 types (83% were typable).

## 2. Phage typing method

In enterococci, no phage-typing method has been standardised so far. The practice remains confined to specialised laboratories. This Section gives the rough principles only. Detailed information may be found in the publications quoted in V.B, 1.

*Phage typing.* The surface of an agar plate is inoculated evenly with a 3–12 h culture of the strain to be typed, grown in Todd-Hewitt broth at 37°C. Excess fluid is discarded. After 15 min (the dishes having been left open to dry) the inoculated plates are spotted with diluted phages using loops (according to Caprioli *et al* (1975) working titres for selected phages varied from $10^4$ to $5 \times 10^9$ plaque-forming units per 1 ml). After 18 h of incubation at 37°C readings are made with a magnifying glass ($\times 10$). Confluent or semiconfluent lysis is recorded.

Untypable strains are grown for 3 h in Todd-Hewitt broth, exposed to 56°C for 2 min and then cooled rapidly before inoculation to increase typability (Hérmán and Hoch, 1971).

*Isolation of phages.* Sewage, faeces or urine are inoculated into Brain Heart Infusion Broth (Difco) and incubated overnight at 37°C. The culture is centrifuged and filtered through a 0·45 $\mu$m Millipore membrane filter. The filtrate is checked for lysis on a set of indicator strains. Further purification of the phage is performed by three or more consecutive single-plaque propagations (Caprioli *et al.*, 1975). In routine typing, the highest dilution of phage giving confluent lysis is used.

*Storage of phages.* Enterococcal phages are quite stable. Undiluted filtrates can be stored for over a year at 4°C, pH 7·0.

## C. Bacteriocin typing

### 1. General principles, enterocinogeny and enterocin sensitivity

The general principles of bacterial classification by means of bacteriocins have been described by Mayr-Harting et al. (1972) in this series, Volume 7A. In enterococci, bacteriocinogeny was first observed by Kjems (1955); enterococcal bacteriocins have been named "enterocins" (Brandis and Brandis, 1962; Brandis et al., 1965). The frequency of production and the activity spectra of enterocins were studied by Pleceas (1970) and Tzannetis et al. (1970) who found enterocinogeny in nearly 80% of enterococcus strains. The rate was 100% in S. faecalis and its varieties; less frequent production was found in S. faecium and S. durans as well as in non-enterococcus group D strains, i.e. S. bovis and S. equinus. Enterocin activity spectra were variable and depended on the set of indicator strains used. The majority of strains were sensitive to one, two, or usually to several enterocins. The sensitivity spectra of individual strains were constant (Pleceas et al., 1972). Enterocinogeny and enterocin-sensitivity may coexist in the same strain. It was generally supposed that a strain is always resistant to its own enterocin (Brock et al., 1963), an analogy with the relation between colicinogeny and colicin-sensitivity (Fredericq, 1956) was assumed. Surprisingly in five strains, susceptibility to their own enterocins was found (Tzannetis et al., 1970).

Enterocins have no inhibiting activity against Gram-negative species, this being consistent with Hamon's (1965) postulate that bacteriocins of Gram-positive bacteria are not active against Gram-negative bacteria and vice versa (Iverson and Millis, 1976).

In contrast to what Mayr-Harting et al. (1972) state concerning bacterio-cinogeny typing in other bacteria, in enterococci the sensitivity spectrum principle appears to afford a more advantageous approach to practical typing (Pleceas et al., 1974). The reason for this is the general sensitivity of strains and long-term stability of their sensitivity spectra. Another reason is rather theoretical: bacteriocinogeny is closely related to the species of the producing strain. Some strains are only weak producers or non-producers and could not be included in an enterocinogeny-typing scheme (Pleceas et al., 1972). By sensitivity testing, on the contrary, not only can nearly all enterococcus strains be differentiated, but even some non-enterococcal group D strains, e.g. S. bovis or S. equinus, which have been found to be sensitive not only to bacteriocins of homologous species (if produced) but also to enterocins produced by other group D organisms (Pleceas et al., 1971).

### 2. Enterocin typing methods

So far, available sets of enterocinogenic or sensitivity indicator strains have not been unified, nor have these strains been accorded generally accepted designations. Furthermore, there is no standard method in use, although its introduction would be valuable, particularly in view of the dependence of bacteriocinogeny on culture medium composition and growth conditions. It may be generally stated here that the most practicable procedure available is the double layer agar plate method of Fredericq (1948) in its various modifications described in the literature and referred to in Section V.C.

The double layer agar plate method is biphasic: broth cultures of enterocin producers are inoculated onto the surface of a bottom agar layer and incubated. Then agar containing a suspension of the broth cultures to be tested is overlaid. After incubation, growth inhibition zones are evaluated.

Another approach is to determine inhibition of a sensitive strain by bacteriocin activity by measuring the optical density of broth cultures during the early log phase at short consecutive time intervals (Schlegel and Slade, 1972; Iverson and Millis, 1976).

The bacteriocin-(enterocin-) typing method is a promising one. It allows differentiation of strains independently of serotyping or phage typing into at least 25 distinct types. It will be able to provide complementary information on individual strains belonging to the same taxon. The method urgently requires further improvement to render it useful in general microbiological diagnostic laboratories.

## REFERENCES

Bauer, H., Farr, D. R., and Horisberger, M. (1974). *Archs Microbiol.*, **97**, 17–26.
Beckerich, A., and Hauduroy, P. (1922). *C. r. hebd. Séanc. Soc. Biol.*, **86**, 881–882.
"Bergey's Manual of Determinative Bacteriology", 8th edn (1974). Williams and Wilkins, Baltimore.
Bleiweis, A. S., and Krasue, R. M. (1965). *J. exp. Med.*, **122**, 237–249.
Bottone, E., Allerhand, J., and Pisano, M. A. (1974). *Antonie van Leeuwenhoek J. Microbiol. Serol.*, **40**, 385–392.
Brandis, H., and Brandis, U. (1962). *Pathologia Microbiol.*, **25**, 632.
Brandis, H., Brandis, U., and van de Loo, W. (1965). *Zbl. Bakt. Paras. Abt. I Orig.* **196**, 331.
Brock, T. D., Peacher, B., and Pierson, D. (1963). *J. Bact.*, **86**, 702.
Brock, T. D., and Davie, J. M. (1963). *J. Bact.*, **86**, 708–712.
Caprioli, T., Zaccour, F., and Kasatiya, S. S. (1975). *J. clin. Microbiol.*, **2**, 311–317.
Chapman, O. (1944). *J. Bact.*, **48**, 113.
Ciuca, M., Baldovin-Agapi, C., Mihalco, F., Beloiu, I., and Caffé, I. (1959). *Arch. Roum. de Pathol. Expér.*, **18**, 519–526.
Colman, G. (1970). Thesis, University of London.
Cowan, S. T., and Steel, K. J. (1974). "Manual for the Identification of Medical Bacteria", 2nd edn. Cambridge University Press, Cambridge.

Daoust, R. A., and Litski, W. (1975). *Appl. Microbiol.*, **29**, 584.

Deibel, R. H., Lake, D. E., and Niven Jr., C. F. (1963). *J. Bact.*, **86**, 1275–1282.

De Moor, C. E. (1963). *Antonie van Leeuwenhoek J. Microbiol. Serol.*, **29**, 272–280.

Drucker, D. B., and Melville, T. H. (1973). *Microbios*, **7**, 117–131.

Duma, R. J., Weinberg, A. N., Medrek, T. F., and Kunz, L. J. (1969). *Medicine*, **48**, 87–127.

Efthymiou, C. J., and Joseph, S. W. (1974). *Appl. Microbiol.*, **28**, 411–416.

Elliott, S. D. (1959). *Nature, Lond.*, Suppl. No. 17, 184, 1342.

Elliott, S. D. (1966). *J. Hyg. (Camb.)*, **64**, 205–212.

Elliott, S. D., Hayward, J., and Liu, T. Y. (1971). *J. exp. Med.*, **133**, 479.

Evans, A. C. (1934). *Public Hlth Rep.*, **49**, 1386–1401.

Evans, A. C., and Chinn, A. L. (1947). *J. Bacteriol.*, **54**, 495.

Facklam, R. R., and Moody, M. D. (1970). *Appl. Microbiol.*, **20**, 245–250.

Facklam, R. R. (1972). *Appl. Microbiol.*, **23**, 1131–1139.

Facklam, R. R. (1973). *Appl. Microbiol.*, **26**, 138–145.

Falkow, S. (1958). *J. clin. Pathol.*, **29**, 598–600.

Fredericq, P. (1948). *Rev. Belge pathol. Méd. exp.*, **19**, (Suppl. 4, I.).

Fredericq, P. (1956). *C. r. hebd. Séanc. Soc. Biol.*, **150**, 1514.

Fuller, A. T. (1938). *Br. J. exp. Pathol.*, **19**, 130–139.

Garland, J. M., Archibald, A. R., and Baddiley, J. (1975). *J. gen. Microbiol.*, **89**, 73–86.

Gooder, H. (1968). *In* "Microbial Protoplasts, Spheroplasts and L-forms", pp. 40–51 (L. B. Guze, Ed). Williams and Wilkins, Baltimore.

Graham, W. C., and Bartley, E. O. (1939). *J. Hyg.*, **39**, 358.

Guthof, O. (1955). *Zbl. Bakteriol. Abt. I Orig.* 164, 160–169.

Hajna, A., and Perry, C. (1943). *Am. J. Publ. Health*, **33**, 550.

Hamon, Y. (1965). *Pathol. Biol.*, **13**, 806–824.

Hartman, P. A., Reinbold, G. W., and Saraswat, D. S. (1966). *Int. J. Syst. Bact.*, **16**, 197–221.

Hay, J. B., Wicken, A. J., and Baddiley, J. (1963). *Biochim. Biophys. Acta*, **71**, 188–190.

Hehre, E. J., and Neill, J. M. (1946). *J. exp. Med.*, **83**, 147.

Heidelberger, M., and Baddiley, J. (1974). *Carbohyd. Res.*, **37**, 5–7.

Hérmán, G., and Hoch, V. (1971). *Acta Microbiol. Acad. Sci. Hung.*, **18**, 101–104.

Hoch, V., and Hérmán, G. (1971). *Acta Microbiol. Acad. Sci. Hung.*, **18**, 95–99.

Hoppes, W. L., and Lerner, P. I. (1974). *Ann. Int. Med.*, **81**, 588–593.

Iverson, W. G., and Millis, N. F. (1976). *Can. J. Microbiol.*, **22**, 1040–1047.

Jacob, A. E., Douglas, G. J., and Hobbs, S. J. (1975). *J. Bact.*, **121**, 863–872.

Jelínková, J., Rýc, M. (1971). *Čs. Epid. Microbiol. Immunol.*, **20**, 121–125.

Jelínková, J., and Mottl, J. (1975). "Working paper WHO Collaborating Centre for Reference and Research on Streptococci", Prague.

Jones, D., Deibel, R. H., and Niven, Jr., C. F. (1964). *J. Bact.*, **88**, 602–610.

Jones, D., Sackin, M. J., and Sneath, P. H. A. (1972). *J. gen. Microbiol.*, **72**, 439–450.

King, J. R., and Gooder, H. (1970). *J. Bact.*, **103**, 692–696.

Kjems, E. (1955). *Acta path. microbiol. scand.*, **36**, 433–440.

Krämer, J., and Brandis, H. (1975). *J. gen. Microbiol.*, **88**, 93–100.

Krämer, J., and Lenz, W. (1975). *Zbl. Bakt. Abt. I Orig. A*, **231**, 421–425.

Krause, R. M. (1972). *In* "Streptococci and Streptococcal Diseases", Ch. 5, pp. 67–74 (L. W. Wannamaker, Ed). Academic Press, London and New York.

Lancefield, R. C. (1933). *J. exp. Med.*, **57**, 571–595.

Maxted, W. R., and Fraser, C. A. M. (1964). *Zbl. Bakt. Parasitenk.*, **2**, Abt., 196, 76–78.

Mayr-Harting, A., Hedges, A. J., and Berkeley, R. C. W. (1972). In "Methods in Microbiology", Vol. **7A**, pp. 315–422 (J. R. Norris and O. O. Ribbons Eds.). Academic Press, London.

McCandless, R. G., Cohen, M., Kalmanson, G. M., and Guze, L. B. (1968). *J. Bact.*, **96**, 1400-1412.

McCandless, R. G., Hensley, T. J., Cohen, M., Kalmanson, G. M., and Guze, L. B. (1972). *J. gen. Microbiol.*, **68**, 357–365.

Mead, G. C. (1963). *Nature, Lond.*, **197**, 1323–1324.

Montgomerie, J. Z., Kalmanson, G. M., Harwick, H. J., and Guze, L. B. (1973). *Proc. Soc. exp. Biol. Med.*, **144**, 868–870.

Mundt, J. O., and Graham, W. F. (1968). *J. Bact.*, **95**, 2005–2009.

Nowlan, S. S., and Deibel, R. H. (1967). *J. Bact.*, **94**, 297–299.

Parker, M. T., and Ball, L. C. (1976). *J. Med. Microbiol.*, **9**, 275–302.

Pavlova, M. (1969). Thesis (part II), Charles University, Prague.

Pavlova, M., Beauvais, E., Brezenski, F. T., and Litski, W. (1972). *Appl. Microbiol.* **23**, 571–577.

Pleceas, P. (1967). *Arch. Roum. Path. exp. Microbiol.*, **26**, 405–410.

Pleceas, P., Racovita, C., Thomas, E., and Epuran, E. (1968). *Arch. Roum. Path. exp. Microbiol.*, **27**, 303–308.

Pleceas, P. (1970). *Arch. Roum. Path. exp. Microbiol.*, **29**, 229–232.

Pleceas, P., Bogdan, C., and Vereanu, A. (1971). *Arch. Roum. Path. exp. Microbiol.*, **30**, 351.

Pleceas, P., Bogdan, C., and Vereanu, A. (1972). *Zbl. Bakt. Hyg., Abt. I Orig. A*, **221**, 173–181.

Pleceas, P., Barbe, G., Bringuies, J. P., and Carraz, M. (1974). *Rev. Inst. Pasteur Lyon*, **7**, 113–117.

Pleceas, P., and Brandis, H. (1974). *J. Med. Microbiol.*, **7**, 529–533.

Pleceas, P., and Brandis, H. (1975). VIth International Symposium on *Streptococcus pyogenes*, Prague, Abstract, p. 46.

Portas, M. R., Hogan, N. A., and Hill, H. R. (1976). *Lab. clin. Med.*, **88**, 339–344.

Sabbaj, J., Suttner, V. L., and Finegold, S. M. (1971). *Appl. Microbiol.*, **22**, 1008–1011.

Schlegel, R., and Slade, H. D. (1972). *J. Bact.*, **112**, 824–829.

Sharpe, M. E., and Shattock, P. M. F. (1952). *J. gen. Microbiol.*, **6**, 150.

Sherman, J. M. (1937). *Bact. Rev.*, **1**, 3–97.

Syerman, J. M. (1938). *J. Bact.*, **35**, 81–93.

Slade, H. D., and Slamp, W. C. (1972). *J. Bact.*, **109**, 691–695.

Smith, D. G., and Shattock, F. P. M. (1964). *J. gen. Microbiol.*, **34**, 165–175.

Smith, H. L., and Goodner, K. (1958). *J. Bact.*, **76**, 662.

Swan, A. (1954). *J. clin. Path.*, **7**, 160–163.

Switzer, R. W., and Evans, J. B. (1974). *Appl. Microbiol.*, **28**, 1086–1087.

Taylor, R. F., Ikawa, M., and Chesbro, W. (1971). *J. Bact.*, **105**, 676–678.

Taylor, R. F., and Davies, B. H. (1974a). *Biochem. J.*, **139**, 751–760.

Taylor, R. F., and Davies, B. H. (1974b). *Biochem. J.*, **139**, 761–769.

Thiercelin, M. E. (1899). *C. r. Soc. Biol.*, **51**, 269–271.

Toon, P., Brown, P. E., and Baddiley, J. (1972). *Biochem. J.*, **127**, 399.

Tzannetis, S., Leonardopoulos, J., and Papavassiliou, J. (1970). *J. appl. Bact.*, **33**, 358–362.

Van der Drift, C., Duiverman, J., Bexkens, H., and Krijnen, A. (1975). *J. Bact.*, **124**, 1142–1147.

Von Graevenitz, A., Redys, J. J., and Cassidy, E. (1970). *J. Med. Microbiol.*, **3**, 227–232.

Watanakunakorn, Ch. (1974). *Am. J. Med.*, **56**, 256–260.

Williams, R. E. O., and Hirch, A. (1950). *J. Hyg.*, **48**, 504-524.

Young, L. S., and Armstrong, D. (1969). *J. Inf. Dis.*, **120**, 281–291.

# CHAPTER X

# Bacteriocins of Streptococci and Bacteriocin Typing

### H. BRANDIS

*Institute of Medical Microbiology and Immunology of the University of Bonn*

## I. INTRODUCTION

Bacteriocin typing of streptococci so far is only in a very preliminary state. The reasons for this may lie in the diversity of inhibitory substances produced by streptococci as well as in the lack of reproducible and standardised methods. The main purpose of the following survey shall be the discussion of observations on bacteriocins produced by streptococci in order to show to what extent possibilities for developing typing schemes may exist.

Inhibition phenomena produced by streptococcal strains have been described by many authors (for review of the older literature see Florey *et al.*, 1959). It must be borne in mind, however, that the mechanisms of inhibition are in many cases not clearly characterised and must not always be attributed to the action of bacteriocins. Malke *et al.* (1974), for instance, reported that some inhibitory phenomena among group A streptococci are caused by the toxic action of hydrogen peroxide. This is also true for some strains of *S. sanguis* (group H) (Holmberg and Hallander, 1973). Wolff and Duncan (1974) described a bactericidal substance produced by group A streptococci which is possibly an inhibitory substance other than a bacteriocin. Among streptococci from milk bactericidal substances are produced which in some ways resemble antibiotics used for therapeutic purposes (e.g. Diplococcin, Nisin).

Many of the described streptococcal bacteriocins have been poorly characterised. Only for reasons of simplification are they referred to as bacteriocins (synonyma: streptocins, streptococcins) in this Chapter.

## II. BACTERIOCINS OF GROUP A STREPTOCOCCI

### A. Frequency of bacteriocin production

The frequency of bacteriocin production (regardless of the exact nature of the inhibitory substances) by strains of group A streptococci is summarised in Table I.

TABLE I

**Bacteriocin production by streptococci of group A**

| Percentage of positive strains | Authors |
|:---:|:---|
| 3·1 | Tagg *et al.* (1971) |
| 17·0 | Sherwood *et al.* (1949) |
| 60·0 | Kuttner (1966) |
| 53·0 | Overturf and Mortimer (1970) |
| 38·0 | Prakash *et al.* (1973) |
| 41·0 | Piatkowski *et al.* (1972) |

According to Kuttner (1966), only few strains seem to be suitable as indicators. No correlation exists between serological type and bacteriocin production (Totolyan and Kolesnichenko, 1971) and no relationship between nephritogenicity of a strain and bacteriocin production as suggested by Kuttner (Overturf and Mortimer, 1970).

Bacteriocins of group A streptococci are difficult to separate from living cells (Sherwood et al., 1949; Kuttner, 1966). For detection of bacteriocin production Kuttner as well as Overturf and Mortimer (1970) used a lawn with a low density of indicator organisms, on which were placed drops of young broth cultures of the strains to be tested. This technique may perhaps account for the observation of Kuttner that bacteriocin production was not always constant. However, as Malke et al. (1974) point out, the possibility is not ruled out that the inhibition phenomena observed by these authors are not due to bacteriocin production.

As already observed by Kuttner (1966), Tagg et al. (1973a) found that the producing strain may be sensitive also to homologous bacteriocin, i.e. bacteriocin produced by the same strain.

## B. Mode of action

Tagg et al. (1971, 1973a, b) described a bacteriocin (streptocin A) produced by strain FF-22. This has a molecular weight of approximately 8000. It is bactericidal, but not lytic for sensitive strains. The lethal effect is temperature dependent. Streptocin A inhibits the synthesis of DNA, RNA, and protein, and prevents the uptake and incorporation of glucose. Degradation of RNA occurs.

Thus, streptocin A seems sufficiently characterised to be considered as a bacteriocin. The determinants of streptocin A-FF22 production can be transduced to group A streptococci of different types. Streptomycin resistance markers are not cotransducible with the bacteriocin determinants (Tagg et al., 1976).

## C. Inhibitory spectrum

Streptocin A has a wide spectrum of activity on different Gram-positive organisms but, not on Gram-negative strains (Tagg et al., 1973b). A portion of strains of S. viridans, S. faecalis, Staphylococcus aureus are inhibited.

## D. Bacteriocin typing possibilities

No typing scheme has been developed. Since only less than 60% of the strains are bacteriocin producers, bacteriocin production does not seem very promising as a marker. No extensive experience exists as to whether bacteriocin sensitivity of group A streptococci can be used for typing purposes. As already mentioned, Kuttner who, however, worked with a limited number of strains, rarely found sensitive strains.

## III. BACTERIOCINS OF GROUP B STREPTOCOCCI

### A. Bacteriocin production

There are only a few reports of bacteriocin production by group B streptococci and a typing scheme does not exist.

Sherwood et al. (1959), Kuttner (1966), Prakash et al. (1973) showed that bacteriocin-like activity can be observed among group B streptococci.

Among 121 strains of human origin, eight were found by Krämer and Brandis (1972) to produce inhibition zones on group B indicator strains. The chloroform stabbing method was used. The medium used consisted of 0·5% NaCl, 0·05% glucose, 0·12% Yeast Extract (DIFCO), 1% Meat Extract (Liebig), 1% Peptone (Merck), 0·02% $CaCl_2$, 5% calf serum and 1·5% agar, pH 7·2–7·3. Only four strains gave inhibition zones with a diameter of more than 10 mm. These four strains could be differentiated from one another by a different host range on 11 group B indicator strains. Ninety-six strains of group B streptococci were sensitive to the bacteriocins produced by these four strains. Only three of the eight bacteriocin producing strains produced a bacteriocin in fluid medium and this occurred only in low titre. A bacteriocin with a high titre was produced by strain S. agalactiae 73. This was characterised and partially purified by Krämer and Brandis (1972). It is a relatively stable non-inducible bacteriocin with a molecular weight of about 10,000. The bacteriocin was sensitive to trypsin and pepsin and inactivated by incubation at 80 °C for 20 min. Tagg et al. (1975) found that two of 105 group B streptococci from human sources produced a bacteriocin. This was referred to as streptocin $B_1$. The two strains belonged to serological type II and had identical inhibitory spectra. The bacteriocin of strain 74-628 was characterised and partially purified. As it was impossible to achieve sufficient bacteriocin production in fluid media, the active substance was extracted by freeze-thawing from cultures grown on Todd-Hewitt Agar. The inhibitory activity was associated with two distinct molecular species. One of these had a molecular weight of about 10,000 and the other probably in excess of 200,000. The crude bacteriocin was stable, non-inducible, and no loss in activity occurred after boiling for 15 min at pH 6·5. It was formed at 35 °C, not at 42 °C, and inactivated by pronase, and to a lesser degree by trypsin.

Tzannetis et al. (1974) confirmed the observations made by Krämer and Brandis (1972) that weak and strong bacteriocin producers occur among group B streptococci. Tzannetis et al. (1974) used Tryptose Phosphate soft agar (0·8%) (DIFCO) as the test medium. The agar plates were stabbed with the test strain and incubated for 48 h at 37 °C and then a layer of soft agar containing the indicator cells was poured over the surface without exposure to chloroform. 45 strains of group B streptococci (22 of human and 23 of bovine origin) were investigated. Thirty strains (66·6%) produced bacteriocins. The frequency of sensitivity to bacteriocins was about the same.

## B. Inhibitory spectrum

Bacteriocin B 73 described by Krämer and Brandis as well as streptocin $B_1$, showed a wide range of activity on a number of strains of Gram-positive, but not of Gram-negative organisms. Streptocin $B_1$ was active on streptococci of group A (98 of 149 strains tested), B (27 of 47 strains), C (7 of 23 strains), D (5 of 29 strains), F, G (3 of 16 strains), *S. salivarius*, *S. pneumoniae*, *Staphylococcus epidermidis*, *Staph. aureus* (3 of 23 strains) and *Micrococcus luteus*.

# IV. BACTERIOCINS OF GROUP H STREPTOCOCCI

Schlegel and Slade (1972) found that group H streptococcus strain Challis produced a bacteriocin streptocin $STH_1$, in Brain Heart Infusion Broth. The bacteriocin was lethal for group H streptococcus strain Wicky. This bacteriocin was sensitive to trypsin, phospholipase C and alkaline phosphatase. It was partially purified by Schlegel and Slade (1973) and appears to exist in two molecular forms (110,000 and 28,000–30,000). The maximum titre was reached after incubation for 120 min, i.e. while the bacteriocin was still in the exponential growth phase. The lethal action of streptocin $STH_1$ followed "one hit" kinetics and was accompanied by inhibition of DNA, RNA, and protein synthesis in Wicky cells. Membrane alterations were also seen (Schlegel and Slade, 1974). The host range of streptocin $STH_1$ has not been determined.

# V. BACTERIOCINS OF *STREPTOCOCCUS MUTANS*

## A. Bacteriocin production

Kelstrup and Gibbons (1969a) demonstrated the production of bacteriocins with six of 13 cariogenic and non-cariogenic strains of humans and rodents. The bacteriocins were active on other streptococcal strains, including *S. pyogenes* and enterococci. Strains of *Lactobacillus*, *Staphylococcus* and *Escherichia coli* remained unaffected. The author used the stab method on Trypticase Soy Agar supplemented with 0·2% glucose and an overlay of soft agar containing the indicator cells. The plates were incubated under aerobic and anaerobic conditions (5% $CO_2$, 10% $H_2$, 85% $N_2$).

The drop assay was also performed. In broth cultures bacteriocin production could be demonstrated by addition of 0·01–0·1% agar or 0·1–1·0% starch only. The bacteriocins were heat stable (80 °C for 45 min with exception of the bacteriocin of strain 130 which was destroyed at 80 °C for 15 min). With the exception of two the bacteriocins were trypsin sensitive and resistant to lipase (except for the bacteriocins of strain PK1 and 130 (partially)). In consequence of their sensitivity to proteolytic enzymes, the bacteriocins of oral streptococci can be destroyed by dental plaque material, saliva and intestinal contents containing proteolytic enzymes (Kelstrup and Gibbons, 1969b). Rogers (1972) tested 15 *S. mutans* strains for bacteriocin production using a *Streptococcus pyogenes* type 12 strain as indicator. The test strains were grown in Todd-Hewitt Broth (Oxoid) and spotted on agar plates with a loop. After an incubation of 24–36 h at 37 °C under an atmosphere of 95% $N_2$ and 5% $CO_2$ a soft agar layer containing the indicator cells was poured onto the surface of the plates. After aerobic incubation for 18 h at 37 °C, the inhibition zones were measured. The medium used had a strong influence on bacteriocin production. This was also stressed by Hamada and Ooshima (1975b) and must be borne in mind if a bacteriocin typing scheme is to be developed. A medium containing 2% Trypticase (BBL), 2·5% NaCl, 4% $H_2HPO_4$ and 1·5% Ionagar No. 2 (Oxoid) with addition of 2% yeast extract (BBL) gave the best results. On this medium 13 of 15 strains showed inhibition of the indicator strain by at least one of the eight producer strains tested. Bacteriocin production of eight strains of *S. mutans* grown on 5% sucrose medium was not influenced. However, when grown in sucrose broth, all five *S. mutans* indicator strains became resistant to the bacteriocins of the eight producer strains due to a coating of extracellular polysaccharides (Rogers, 1974). These findings could not be confirmed, however, by Delisle (1976).

The occurrence of mutacinogenic strains is frequent. Hamada and Ooshima (1975b) applying the stab culture technique (Tryptose Soy medium with 2% agar) to 130 clinical isolates of *S. mutans* from children found 84 strains (74%) which inhibited at least one of the ten indicator strains used (five *S. mutans*, two *S. salivarius*, two *S. sanguis*, one *S. pyogenes*). Among 85 type c strains, 65 were mutacinogenic, among 13 type d strains nine, and among 15 type e/untypable strains ten. *S. pyogenes* NCTC 8198, *S. salivarius* HHT and *S. sanguis* strains ATCC 10556 and OMZ9 proved to be sensitive indicator strains. The mutacinogeny and sensitivity of strains were stable characters as the results were essentially the same throughout a period of 2 years.

The findings of Hamada and Ooshima (1975b) are in good agreement with those by Rogers (1976) who found 70% of 143 strains of *S. mutans* to be active bacteriocin producers towards one or more of a series of indicator strains.

A bacteriocin produced by *S. mutans* GS-5, serotype c, was obtained from cell-free culture fluid (PYG medium, supplemented with 5% horse serum) and was partially purified (Paul and Slade, 1975). It was sensitive to trypsin and pronase and resistant to catalase and to heating at 100°C for 10 min. The molecular weight was more than 20,000. The bacteriocin exerted a bactericidal action on susceptible cells without being bacteriolytic.

In a sterile-filtered fluid medium (one-halfstrength APT broth (BBL) containing 4% (w/v) yeast extract) Delisle (1975) also obtained bacteriocin production by two strains of *S. mutans* (BHT and GS-5). Induction with UV or Mitomycin C was unsuccessful.

It seems that many strains of *S. mutans* produce more than one type of bacteriocin and that low and high molecular weight type bacteriocins are produced by several strains of S. *mutans* (Rogers, 1976). The exact chemical nature and the mode of action, however, are at present unknown.

## B. Inhibitory spectrum

Hamada and Ooshima (1975a) tested 220 *S. mutans* strains against ten indicator strains (five *S. mutans*, two *S. salivarius*, two *S. sanguis* and one *S. pyogenes*). About 75% of the *S. mutans* strains inhibited at least one of the ten indicators. The inhibitory substance was called mutacin. This mutacin action was not due to hydrogen peroxide formation. For detection of mutacin activity the authors used the stab method. Following incubation of the cultures in Trypticase Soy Agar (BBL) for 48 h at 37°C, they were covered with an overlay of soft agar containing the indicator cells (0·5 ml overnight Trypticase Soy Broth (BBL) culture (about $10^7$ colony forming units) in 4 ml melted soft agar). After overnight incubation, results were read and the radius of the inhibition zones measured. Strains of many different species of Gram-positive bacteria were inhibited (e.g. *S. sanguis*, *S. salivarius*, *S. pneumoniae*, *S. lactis*, *S. faecalis*, *S. pyogenes* A, streptococci of other serological groups, *Lactobacillus*, *Micrococcus lysodeikticus*, *Staph. aureus*, *Staph. epidermidis*, *Bacillus megaterium*, *Mycobacterium phlei* and other mycobacteria, *Nocardia*, *Actinomyces viscosus* and others). All tested strains of Gram-negative bacteria were unaffected. Rogers (1976) working with 11 bacteriocinogenic *S. mutans* strains (technique see above, Rogers 1972, 1974) found that the bacteriocin was active not only against *S. mutans* indicator strains, but also against one strain of *S. pyogenes* A-12, *S. faecalis*, *S. salivarius*, *M. luteus*, *Staph. aureus*, *Staph. epidermidis*, *Listeria* sp., *Corynebacterium* sp. and *Bacillus cereus*.

Contrary to the above mentioned findings the bacteriocin of *S. mutans*

GS-5 described by Paul and Slade (1975) inhibited only some strains of *Streptococcus* within serogroups A, C, D, F, G, H, L, and O. *S. mutans* belonging to serotypes a, b, c, and d were unaffected.

Yamamoto *et al.* (1975) carried out cross-tests between 17 strains of oral streptococci (12 *S. mutans*, three S. *sanguis*, one *S. mitis*, and one *Streptococcus* sp.) using the stab culture method on Trypto-Soy-Broth (3%, Eiken, Japan) with 2% agar (DIFCO) and soft agar overlay containing the indicator cells as well as the punch-hole method (Mayr-Harting, 1964). The plate assay media inoculated with the donor strains were incubated at 37°C for 40 h and then overlayed (0·5 ml of an overnight broth culture of the indicator strains in 4 ml melted soft agar). The plates were read after overnight incubation. For the broth assay, several reservoirs were punched out in the agar and filled with the 2–5 fold concentrated supernatant of the test cultures adjusted to pH 7 and sterilised by filtration through a membrane filter (0·45 $\mu$m). The plates were left at 15°C for several hours and were then overlayed with soft agar containing the indicator cells. The bacteriocin of the "rough type" of strain GS-5 showed an inhibition of 7 *S. mutans* and 3 *S. sanguis* strains. Five other *S. mutans* strains also in the "rough type" were only active on *S. sanguis* indicators. Variants of the "mucoid type" of *S. sanguis* did not produce any inhibitory agent. On the contrary, only "mucoid type" variants of *S. mutans* were susceptible to the bacteriocins produced by "rough type" variants. The bacteriocin-like substances were completely inactivated by treatment with Pronase E and Nagarase and were heat stable.

## C. Typing of oral streptococci

Kelstrup *et al.* (1970) developed a method of bacteriocin "fingerprinting" of oral streptococci for epidemiological and ecological studies. The authors tested bacteriocin production against ten selected indicator strains and bacteriocin sensitivity to ten known bacteriocinogenic strains. The agar stab method (Trypticase Soy Broth (BBL) supplemented with 0·2% glucose and 2% Bacto-agar (DIFCO)) with a multiple stab device was used (ten brass needles were simultaneously dipped into broth cultures of the donor strains and then stabbed into the agar plates). The cultures were incubated for 48 h at 35°C in anaerobic jars (5% $CO_2$, 10% $H_2$, 85% $N_2$) and were then overlayed with soft agar containing the indicator cells (2 ml soft agar mixed with 0·5 ml of broth culture of the indicator strain) and incubated 24 h under the same conditions (indicator strains: *Streptococcus* Z, Cl, C4, C7, J4, J8, *S. salivarius* SS$_2$, *S. salivarius* 1A, enterococcus X21, enterococcus X38; producer strains: *S. mutans* G8, H34, 2M W, PK1, FA1, 130, GF71, BHT, *Streptococcus* PK6). The size and shape of the inhibition zones were recorded. The authors state that their method showed a good degree of reproducibility and allowed them to type and distinguish streptococci. The method can be used for ecological studies. However, it is time-consuming and laborious.

Berkowitz and Jordan (1975) used the techniques and media described by Kelstrup and Gibbons (1969a) to test 120 *S. mutans* isolates from four mother-infant pairs in connection with the likelihood of transmission of *S. mutans* strains from mother to infant. 18 indicator strains (nine *S. mutans*, one *S. mitis*, three *Streptococcus*, two *S. sanguis*, two enterococcus, and one *S. salivarius*) were employed. Among 120 *S. mutans* strains 119 produced bacteriocins which acted at least on one of the 18 indicator strains. These indicator strains were designated with a code letter (e.g. A, B, C, and so on). Thus a bacteriocin code could be established for each bacteriocin producing strain based on the pattern of the indicator strains growth inhibition.

A standardised bacteriocin typing scheme for *S. mutans* does not yet exist, but the high incidence among *S. mutans* strains, especially type c, seems to be a good base for developing such a scheme (Rogers, 1976).


## VI. BACTERIOCINS OF THE *STREPTOCOCCUS VIRIDANS* GROUP

Gerasimov (1968) as well as Piatkowski and Szropińska (1971) investigated the bacteriocinogeny of strains of the *S. viridans* group. Piatkowski and Szropińska used nutrient agar supplemented with 5% defibrinated calf blood or enriched with 10% horse serum. The strains to be tested were inoculated as a point. After incubation for 24 h at 37°C the surface of the medium was sprayed with young broth cultures of the indicator cells in the exponential growth phase. The study comprised 215 strains.

| Strain | Number of strains examined | Number of active strains |
|---|---|---|
| *S. salivarius* | 76 | 44 |
| *S. mitis* | 87 | 54 |
| *Streptococcus* s.b.e. | 13 | 9 |
| *Streptococcus* M.G. | 9 | 4 |
| Undefined | 30 | 4 |
| Total | 215 | 115 (53%) |

With the aid of seven indicator strains ten tentative bacteriocin types could be distinguished (Table II)

TABLE II

**Spectrum of activity of 115 bacteriocinogenic strains of the *S. viridans* group on seven indicator strains (three strains of *S. mitis*, two *S. salivarius*, two not classifiable) according to Piatkowski and Szropińska (1971)**

| Type description | Indicator strains | | | | | | | Number of strains |
|---|---|---|---|---|---|---|---|---|
|  | 1 | 2 | 3 | 4 | 5 | 6 | 7 |  |
| 1 | + | + | + | + | + | + | + | 16 |
| 2 | + | + | + | + | + | − | − | 8 |
| 3 | + | + | + | − | − | − | − | 5 |
| 4 | + | + | + | − | + | − | + | 6 |
| 5 | − | − | + | + | + | + | − | 21 |
| 6 | − | + | − | + | + | − | + | 16 |
| 7 | − | + | + | + | − | + | + | 18 |
| 8 | − | − | + | + | − | − | − | 5 |
| 9 | − | − | − | − | + | − | − | 11 |
| 10 | − | − | + | − | − | − | − | 9 |

The stability of these types and the possibility of using such a scheme for epidemiological or ecological investigations were not considered by the authors.

Dajani *et al.* (1976) detected bacteriocin-like activities in 78% of alpha-haemolytic streptococci. The active substances were called viridins. For detection the following methods were used (a) the simultaneous antagonism technique. The indicator strains were seeded on Tryptic Soy agar plates and then the producer strains were spotted on the plates. After incubation at 37°C overnight, the occurrence of inhibitory zones around the growth of the producer strains was recorded. (b) the deferred antagonism method. Tryptic Soy agar plates were inoculated with spots of the producer strains. After incubation for 24–48 h, the producer strains were killed with chloroform vapour (30 min) and then overlayed with soft agar (10 ml) containing the indicator cells. Screening for zones of inhibition was performed after incubation for 24 h at 37°C. Three strains (*S. sanguis* 24658, *S. mitis* 42885, and *S. mitis* 42991) of viridans streptococci were selected as producer strains to test the spectra of activity. The inhibitory substances of these three strains were called viridin A, viridin B and viridin C, respectively. These viridins showed in the deferred antagonism technique a wide range of activity not only against Gram-positive bacteria, like group

A streptococci, group B streptococci, viridans streptococci, coagulase-negative staphylococci, *S. aureus*, but also against Gram-negative species, like *Neisseria gonorrhoeae*, *Neisseria* sp., *Pseudomonas*, *E. coli*, *Shigella*, *Salmonella*. This activity spectrum is in contrast to those of other bacteriocins of Gram-positive bacteria. The inhibitory substances could not be obtained in fluid media. However, by disruption of streptococcal cells, the active substance could be recovered. Viridin B was partially purified. It had some unusual properties. It was destroyed by trypsin and protease and by heat (65 °C for 30 min). With four sensitive strains tested no adsorption of viridin B to the susceptible cells could be detected. Viridin B was bacteriocidal against *Neisseria sicca*. Against a coagulase-negative staphylococcus strain, however, it was only bacteriostatic.

## VII. BACTERIOCINS OF GROUP D STREPTOCOCCI
(synonyms: enterocins, enterococcins, D streptococcins, D streptocins)

The bacteriocins produced by enterococci are the most intensively investigated group of streptocins. There is no doubt, however, that these D streptocins comprise very heterogeneous bactericidal substances: defective phages, bactericidal substances of relatively low molecular weight and bacteriolysins.

### A. Bacteriocins with phage-like structures

Bradley (1967) described several kinds of particles resembling phage structures in a Mitomycin C induced lysate of strain *S. faecalis* NCIB 8256 (NCTC 8175) which produces a bacteriocin (Brock *et al.*, 1963). Most frequent were particles resembling coliphage T3 with short non-contractile tails showing full or empty heads. Next in frequency were particles which resembled headless tails (about 1000 Å long and 70 Å thick). According to Bradley they resemble a phage tail, but "it is quite different from anything described so far". Furthermore the lysate contains, apart from hollow tubes, a small number of intact phage-like particles with contractile sheaths and empty heads.

### B. Bacteriocins with relatively low molecular weight

The bacteriocin produced by *S. faecalis* E1 which has been described by Brandis and Brandis (1962) and Brandis *et al.* (1965) does not with certainty resemble any phage structures. This bacteriocin, designated enterocin E1 has been partially purified and characterised by Krämer and Brandis (1975a) and its mode of action has been studied (Krämer and Brandis, 1975b). Enterocin E1 occurs in two molecular forms, enterocin E1A and enterocin E1B. Enterocin E1A is a basic substance with a molecular weight of about 10,000. It is resistant to heat and is trypsin sensitive. It represents more than 90% of the total activity of the supernatant fluid of Tryptose Phosphate Broth cultures. Enterocin E1B with a particle weight greater than $4 \times 10^6$ was the predominant type in extracts of bacterial cells. Both forms of purified enterocins have nearly the same activity spectrum and act on certain strains of enterococci, *S. salivarius* and *L. monocytogenes*, but not on Gram-negative bacteria. Susceptible cells are rapidly killed by the enterocins E1A and E1B, but lysis does not occur. The bacteriocins inhibit protein sythesis and drastically reduce the biosynthesis of DNA and RNA, but cause no cell disintegration. E1A also produces membrane alterations.

## C. Bacteriolysins

Finally, bacteriolysins form a third group of bactericidal substances produced by enterococci. The strains *S. faecalis* subsp. *zymogenes* X14 described by Brock *et al.* (1963) produce a bacteriocin which seems to be identical to the haemolysin (Brock and Davie, 1963; Davie and Brock, 1966; Basinger and Jackson, 1968; Granato and Jackson, 1969; Jackson, 1971; Granato and Jackson, 1971a, b; Appelbaum and Zimmerman, 1974). This bacteriocin is active against all enterococci except *S. faecalis* subsp. *zymogenes* and inhibits strains of several other Gram-positive species, but not Gram-negative bacteria (Brock *et al.*, 1963; Jackson, 1971).

The occurrence of bacteriolysins with specific activity is not restricted to *S. faecalis* subsp. *zymogenes*. Brandis and Brandis (1966) described two bacteriolysins of *S. faecalis* subsp. *liquefaciens* which were bacteriocin-like especially in host range and appearance of inhibition zones. These lysins which were, however, non-haemolytic acted on all 26 *S. faecium* strains tested and on four of 26 *S. faecalis* strains but 21 strains of *S. faecalis* subsp. *liquefaciens* remained unaffected.

The bacteriocin derived from *S. faecalis* subsp. *zymogenes* strain E-1 has a pronounced selective activity against *S. pneumoniae* and non-haemolytic enterococci. No activity was found against beta-haemolytic streptococci, non-haemolytic streptococci, beta-haemolytic enterococci, staphylococci and other Gram-positive bacteria, (Bottone *et al.*, 1971). The bacteriocin was inactivated after 20 min at 80°C and resistant to trypsin. It was also active against *Clostridium perfringens* and appears to be bactericidal and only partially bacteriolytic against those bacteria (Bottone *et al.*, 1974).

## D. Plasmids determining bacteriocin and haemolysin production

Tomura *et al.* (1973) observed that transmission of bacteriocinogenicity by conjugation occurred when a bacteriocin producing *S. faecalis* subsp.

*liquefaciens* strain (Streptomycin (SM)- and Tetracyclin (TC)-resistant, bacteriocin I (bact-I) and haemolysin (haem) producer) was grown in broth together with a non-producer also belonging to *S. faecalis* subsp. *liquefaciens* (Penicillin (PC)-resistant, bact-I⁻, haem⁻). Recombinants (PCʳ, SMˢ, TCˢ, Bact-I⁺, haem⁺) were obtained with high frequency. The transfer of haem⁺ was always correlated with the acquisition of bacteriocinogenicity.

The plasmid nature of resistance markers and haemolysin/bacteriocin determinants of *S. faecalis* strain DS-5 was shown by Dunny and Clewell (1975). The plasmid pAMα1 determined Tetracyclin resistance and the plasmid pAMβ1 Erythromycin and Lincomycin resistance. The third plasmid pAMγ1 coded for haemolysin and bacteriocin production. This plasmid was able to promote its own transfer to a plasmid-free *S. faecalis* recipient and could mobilise the non-transferable plasmid pAMα1. The plasmid pAMγ1 had a molecular weight of $34 \times 10^6$ and the number of copies per chromosomal genome has been calculated to be 5·3. These findings were confirmed by Jacob *et al.* (1975) who worked with *S. faecalis* subsp. *zymogenes* strains JH1 and JH3. Both strains showed beta-haemolytic activity which was associated with bacteriocin activity and was coded by plasmids. The plasmids of both strains were self-transferable by conjugation to other *S. faecalis* strains in mixed broth cultures with high frequency. Thus, the bacteriocin/haemolysin production of *S. faecalis* subsp. *zymogenes* and *S. faecalis* subsp. *liquefaciens* at least in the strains investigated was plasmid-coded.

### E. D streptocin production

Many strains of enterococci produce inhibitory substances which will be designated D streptocins in the following for reasons of simplification, although they undoubtedly represent several different phenomena.

D streptocin production is observed by strains of all species and subspecies of group D streptococci, varying in frequency between 26 and 100% (Brock *et al.*, 1963; Brandis *et al.*, 1965; Pleceas 1970a; Tzannetis *et al.*, 1970). The host range of activity of the D streptocins on enterococci varies. Brock *et al.* (1963) distinguished five types of D streptocins. At least some D streptocins in many respects resemble colicins (Brandis *et al.*, 1965) and a classification can be performed using D streptocin resistant mutants (Brandis and Brandis, 1962). For detection of D streptocin production Brandis *et al.* (1965) used meat extract agar supplemented with yeast extract and 5% calf serum (as already described under Section III, Krämer and Brandis, 1972). The plates were inoculated with overnight Trytose Phosphate Broth (DIFCO) cultures. After incubation for 24–48 h at 37°C the macrocolonies were killed by chloroform vapour (25 min) and the plates were then flooded with an overnight Trypticase Phosphate Broth culture of the indicator strain. Reading was performed after incubation overnight. Brock *et al.* (1963) inoculated Todd-Hewitt agar (DIFCO) from overnight broth (Todd-Hewitt) cultures by loop (five or six strains per plate). After incubation for 18-24 h at 37 °C the plates were overlayed with 2 ml of 0·75% agar to which 0·1 ml of the indicator broth culture in the exponential growth phase had been added. Following an incubation of 6–8 h at 37 °C the plates were checked for zones of inhibition. Tzannetis *et al.* (1970) used the double layer method like Brock *et al.* (1963). The strains were grown on the following media: (a) Todd-Hewitt Broth (Oxoid), (b) nutrient agar containing 2% (w/v) bacteriological peptone (Oxoid), 1% (w/v) of dextrose and 0·8% Ionagar No. 2 (Oxoid), (c) soft agar consisting of Nutrient Broth (Oxoid) with 0·7% (w/v) Agar No. 3 (Oxoid). Plates were inoculated with 48 h broth cultures and were then incubated for 48 h at 37 °C. Then an overlay of soft agar (1 ml of 24 h broth cultures of the indicator strains mixed with 7 ml of melted soft agar) was poured onto the plates which were further incubated for 18 h at 37 °C. Pleceas (1970a) used nutrient agar (Merck) supplemented with 5% horse serum. The strains were grown in broth (Institut Cantacuzino, see Pleceas 1970a) supplemented with 2% glucose. The test strain grown in broth for 18–24 h at 37 °C was inoculated with a loop as a spot on the surface of the agar plate. After drying without any incubation a soft agar overlay (2–3 ml of a 24 h broth culture of the indicator strain, diluted 1:100) was poured onto the plates which were then incubated for 24 h at 37 °C before reading the zones of inhibition which appeared around the growth spots of the donor strains.

### F. Inhibitory spectrum

D streptocins show a wide range of activity against strains of different species and subspecies of D streptococci, streptococci of different serogroups, viridans streptococci, staphylococci, *Corynebacterium diphtheriae* and are especially active on *L. monocytogenes* (Brandis and Brandis, 1962; Brock *et al.*, 1963; Pleceas *et al.*, 1971; Krämer and Brandis, 1975a). Remarkable is the high activity of D streptocinogenic strains against *Clostridium perfringens and Cl. septicum.* Of 21 D streptocinogenic strains of *S. faecium, S. faecalis, S. faecalis* subsp. *zymogenes, S. faecalis* subsp. *liquefaciens* 14 strains were found to produce at least one bacteriocin that was active against strains of clostridia. According to their activity spectra several types of D streptocins could be differentiated (Krämer and Schallehn, 1974). In 1960, Stark reported on the high incidence of inhibition of *C. perfringens* by three strains of *S. faecalis* subsp. *zymogenes*. The inhibition of clostridia by enterococci was also noted by Kafel and

Ayres (1969) and Pleceas *et al.* (1971). However it is only with characterised and purified substances that a true impression of the activity spectrum of D streptocins can be made (Krämer and Brandis, 1974a).

## G. D streptocin typing

With the method described earlier Pleceas *et al.* (1972) investigated the sensitivity of 471 strains of group D streptococci against a set of 12 known D streptocinogenic bacteriocin producer strains. These producer strains consisted of two *S. faecalis* subsp. *zymogenes*, six *S. faecalis* subsp. *liquefaciens*, two *S. faecalis* and two *S. faecium* strains. 92·2% of the strains tested were sensitive to one or several D streptocins. The sensitive strains could be classified according to their reactions into 25 types. Type I occurred in 30·1% and Type II in 11% of the strains. 3·8% of the strains were resistant and 6·7% gave diverse results, each characteristic for one strain only. Pleceas *et al.* (1972) suggest that D streptocin typing may be of value for ecological and epidemiological purposes. However, in a number of cases the occurrence of very small inhibition zones make a clear differentiation between positive and negative reactions difficult (Brandis and Pleceas, unpublished results) and the method must be improved before it can be used for routine typing purposes (Pleceas *et al.*, 1974).

Kékessy and Piguet (1971) made cross-tests with 81 strains of *S. faecalis* serving as both producers and indicator strains. The method used was as follows. The strains to be tested as bacteriocin producers were grown in Nutrient Broth (Oxoid) supplemented with 1% glucose for 4–6 h at 37 °C. Then agar plates (DST-Agar (Oxoid) with 1·2% agar) were inoculated with the producer strains by stabbing. For this purpose a multiple inoculator was used. Twenty-seven different strains could be inoculated simultaneously on to one plate by the inoculator. After 48 h at 37 °C, the agar was inverted to a new plastic plate. Thereafter the surface was flooded with a 6–8 h old broth culture which was to serve as indicator of bacteriocin production. After incubation for 8 h at 37 °C, the inhibition zones were recorded (Kékessy and Piguet, 1970). This method avoids the difficulty that chloroform vapour cannot be used with plastic plates, since it dissolves plastic. A further advantage is that phage particles cannot diffuse through the agar and that the inhibition zones observed therefore are not due to bacteriophage action.

Kékessy and Piguet (1971) selected six enterocinogenic strains with different inhibition spectra as indicator strains to monitor the sensitivity of a strain to be tested (enterocino-typing) and six especially sensitive strains as indicators for streptocin production by a strain to be typed (enterocinogeno-typing). The technique of sensitivity testing differed from the method described before in the way of inoculation. The strain to be typed was inoculated as a streak (5 mm wide) across the plate. After incubation for 48 h at 37 °C, the six known indicator strains, after transfer of the inverted agar to a new plate, were inoculated transversely to the original streaks of the strains to be typed.

Among 130 strains of *S. faecalis* typed by Kékessy and Piguet (1971) about 60% (77 strains) were D streptocinogenic whereas all 130 strains were sensitive against D streptocins of one or more of the six known producer strains. It was possible to distinguish 13 types of different sensitivity patterns and ten types of different D streptocin production. By combining both methods a total of 25 types was observed. Type 1, which showed no D streptocin production and was sensitive to two strepto-cinogenic strains, occurred in 25 of 130 strains. The frequencies of some other more common types were as follows: type 2 in ten strains, type 3 in 12 strains, type 4 in eight strains.

## H. Assessment of the typing methods

As Kékessy and Piguet (1971) pointed out, the bacteriocin typing method could be useful for epidemiological purposes, but no trials have been done in this direction. Also, investigations on the reproducibility of the typing results are lacking. If the types were practically constant and reproducible, the method could be applied to investigate the ecology and epidemiology of group D streptococci. It would be interesting to know for example, if special types are connected with urinary tract or other infections. Bacteriocin typing could be useful for revealing cross-infections in hospitals. The distribution of types in food especially in cases of food poisoning would also be informative. Montgomerie *et al.* (1973) observed a possible association between bacteriocin production by four *S. faecalis* subsp. *liquefaciens* strains and their ability to produce a haematogenous pyelone-phritis in the rat. Pleceas *et al.* (1972) found bacteriocin sensitivity to be a stable character of enterococci. The same reactions were found by repeated typing after 1 year. No relations could be detected between D streptocin type and serological or phage type but extensive data to elucidate the question do not exist.

At present, it seems that the serological typing of enterococci (Sharpe and Fewins, 1960; Sharpe, 1964) and the phage typing of enterococci (Pleceas, 1969; Hérmán and Hoch, 1971; Hoch and Hérmán, 1971; Pleceas *et al.*, 1974; Pleceas and Brandis, 1974a, b; Caprioli *et al.*, 1975) have been studied more extensively than bacteriocin typing which is still in a very preliminary state. However the efforts to develop a bacteriocin typing scheme of streptococci for epidemiological purposes have been promising and should be continued.

## REFERENCES

Appelbaum, B., and Zimmerman, L. N. (1974). *Infect. Immunity*, **10**, 991–995.
Basinger, S. F., and Jackson, R. W. (1968). *J. Bact.*, **96**, 1895–1902.
Berkowitz, R. J., and H. V. Jordan (1975). *Archs. oral Biol.*, **20**, 725–730.
Bottone, E., Allerhand, J., and Pisano, M. A. (1971). *Appl. Microbiol.*, **22**, 200–204.
Bottone, E., Allerhand, J., and Pisano, M. A. (1974). *Antonie van Leeuwenhoek*, **40**, 385–392.
Bradley, D. E. (1967). *Bact. Rev.*, **31**, 230–314.
Brandis, H., and Brandis, U. (1962). *Path. Microbiol.*, **26**, 688–695.
Brandis, H., Brandis, U., and van de Loo, W. (1965). *Zbl. Bakt. I Orig.*, **196**, 331–340.
Brandis, H., and Brandis, U. (1966). *Zbl. Bakt. I. Orig.*, **200**, 8–12.
Brock, T. D., and Davie, J. M. (1963). *J. Bact.*, **86**, 708–712.
Brock, T. D., Peacher, B., and Pierson, D. (1963). *J. Bact.*, **86**, 702–707.
Caprioli, T., Zaccour, F., and Kasatiya, S. S. (1975). *J. clin. Microbiol.*, **2**, 311–317.
Dajani, A. S., Tom, M. C., and Law, D. J. (1976). *Antimicrob. Agents Chemother.*, **9**, 81–88.
Davie, J. M., and Brock, T. D. (1966). *J. Bact.*, **92**, 1623–1631.
Delisle, A. (1975). *Antimicrob. Agents Chemother.*, **8**, 707–712.
Delisle, A. L. (1976). *Infect. Immun.*, **13**, 619–626.
Dunny, G. M., and Clewell, D. B. (1975). *J. Bact.*, **124**, 784–790.
Florey, H. W., Chain, E., Heatley, N. G., Jennings, M. A., Sanders, A. G., Abraham, E. P., and Florey, M. E. (1949). "Antibiotics", Vol. 1, pp. 499–506. Oxford Medical Publications,
Gerasimov, A. V. (1968). *Zh. Mikrobiol. Epidem. Immunbiol.*, **45**, 28–32 (in Russian).
Granato, P. A., and Jackson, R. W. (1969). *J. Bact.*, **100**, 865–868.
Granato, P. A., and Jackson, R. W. (1971a). *J. Bact.*, **108**, 804–808.
Granato, P. A., and Jackson, R. W. (1971b). *J. Bact.*, **107**, 551–556.
Hamada, S., and Ooshima, T. (1975a). *J. Dental Res.*, **54**, 140–145.
Hamada, S., and Ooshima, T. (1975b). *Archs. oral Biol.*, **20**, 641–648.
Hérmán, G., and Hoch, V. (1971). *Acta Microbiol. Hung.*, **18**, 101–104.
Hoch, V., and Hérmán, G. (1971). *Acta Microbiol. Hung.*, **18**, 95–99.
Holmberg, K., and Hallander, H. O. (1973). *Archs. oral Biol.*, **18**, 423–434.
Jackson, R. W. (1971). *J. Bact.*, **105**, 156–159.
Jacob, A. E., Douglas, G. J., and Hobbs, S. J. (1975). *J. Bact.*, **121**, 863–872.
Kafel, C., and Ayres, J. C. (1969). *J. appl. Microb.*, **32**, 217–232.
Kékessy, D. A., and Piguet, J. D. (1970). *Appl. Microb.*, **20**, 282–283.
Kékessy, D. A., and Piguet, J. D. (1971). *Path. Microbiol.*, **37**, 113–121.
Kelstrup, J., and Gibbons, R. J. (1969a). *Archs. oral Biol.*, **14**, 251–258.
Kelstrup, J., and Gibbons, R. J. (1969b). *J. Bact.*, **99**, 888–890.
Kelstrup, J., Richmond, S., West, C., and Gibbons, R. J. (1970). *Archs. oral Biol.*, **15**, 1109–1116.
Krämer, J., and Brandis, H. (1972). *Zbl. Bakt. I Orig., A*, **219**, 290–301.
Krämer, J., and Brandis, H. (1975a). *J. gen. Microbiol.*, **88**, 93–100.
Krämer, J., and Brandis, H. (1975b). *Antimicrob. Agents Chemother.*, **7**, 117–120.
Krämer, J., and Schallehn, G. (1974). *Zbl. Bakt. I Orig.*, **226**, 105–113.
Kuttner, A. (1966). *J. exp. Med.*, **124**, 279–291.
Malke, H., Starke, R., Jacob, H. E., and Köhler, W. (1974). *J. med. Microbiol.*, **7**, 367–374.

Mayr-Harting, A. (1964). *J. Path. Bact.*, **87**, 255–266.

Montgomerie, J. Z., Kalmanson, G. M., Harwick, H. J., and Guze, L. B. (1973). *Proc. soc. exp. Biol. Med.*, **144**, 868–870.

Overturf, G. D., and Mortimer, E. A. (1970). *J. exp. Med.*, **132**, 694–701.

Paul, D., and Slade, H. D. (1975). *Infect. Immunity*, **12**, 1375–1385.

Piatkowski, K., Ulewicz, K., Szydlowska, T., and Grzybowski, A. (1972). *Ann. Immunol. Hung.*, **16**, 337–349.

Piatkowski, K., and Szropińska, D. (1971). *Arch. Immunol. Ther. éxp.*, **19**, 137–145.

Pleceas, P. (1969). *Arch. Roum. Path. Exp. Microbiol.*, **28**, 1027–1032.

Pleceas, P. (1970a). *Arch. Roum. Path. Exp. Microbiol.*, **29**, 229–232.

Pleceas, P. (1970b). *Zbl. Bakt. I. Abt. Orig.*, **214**, 130–143.

Pleceas, P., Barbe, G., Bringuier, J. P. and Carraz, M. (1974). *Rev. Inst. Pasteur Lyon*, **7**, 113–117.

Pleceas, P., and Brandis, H. (1974a). *Ann. Microbiol. (Inst. Pasteur)*, **125 B**, 463–470.

Pleceas, P., and Brandis, H. (1974b). *J. med. Microbiol.*, **7**, 529–533.

Pleceas, P., Bogdan, C., and Vereanu, A. (1971). *Arch. Roum. Path. Exp. Microbiol.*, **30**, 351–358.

Pleceas, P., Bogdan, C., and Vereanu, A. (1972). *Zbl. Bakt. I Orig. A*, **221**, 173–181.

Prakash, K., Ravindran, P. C., and Sharma, K. B. (1973). *Ind. J. Med. Res.*, **61**, 1261–1264.

Rogers, A. H. (1972). *Appl. Microbiol.*, **24**, 294–295.

Rogers, A. H. (1974). *Antimicrob. Agents Chemother.*, **6**, 547–550.

Rogers, A. H. (1976). *Archs. oral Biol.*, **21**, 99–104.

Schlegel, R., and Slade, H. D. (1972). *J. Bact.*, **112**, 824–829.

Schlegel, R., and Slade, H. D. (1973). *J. Bact.*, **115**, 655–661.

Schlegel, R., and Slade, H. D. (1974). *J. gen. Microbiol.*, **81**, 275–277.

Sharpe, E., and Fewins, B. G. (1960). *J. gen. Microbiol.*, **23**, 621–630.

Sharpe, E. (1964). *J. gen. Microbiol.*, **36**, 151–160.

Sherwood, N. P., Russell, B. E., Jay, A. R., and Bowman, K. (1949). *J. infect. Dis.*, **84**, 88–91.

Stark, J. M. (1960). *Lancet*, **I**, 733–734.

Tagg, J. R., Dajani, A. S., Wannamaker, J. W., and Gray, E. D. (1973a). *J. exp. Med.*, **138**, 1168–1183.

Tagg, J. R., Dajani, A. S., and Wannamaker, L. W. (1975). *Antimicrob. Agents Chemother.*, **7**, 764–772.

Tagg, J. R., Pihl, E. A., and McGiven, A. R. (1973b). *J. gen. Microbiol.*, **79**, 167–169.

Tagg, J. R., Read, R. S. D., and McGiven, A. R. (1971). *Pathology*, **3**, 277–278.

Tagg, J. R., Read, R. S. D., and McGiven, A. R. (1973c). *Antimicrob. Agents Chemother.*, **4**, 214–221.

Tagg, J. R., Skjold, S., and Wannamaker, L. W. (1976). *J. exp. Med.*, **143**, 1540–1544.

Tomura, T., Hirano, T., Ito, T., and Yoshioka, M. (1973). *Jap. J. Microbiol.*, **17**, 445–452.

Totolyan, A. A., and Kolesnichenko, T. G. (1971). *Zh. Microbiol., Epidem., Immunbiol.*, No. 8, 70–73 (in Russian).

Tzannetis, S., Leonardopoulos, J., and Papavassiliou, J. (1970). *J. appl. Bact.*, **33**, 358–362.

Tzannetis, S., Poulaki-Tsontou, A., and Papavassiliou, J. (1974). *Path. Microbiol.*, **41,** 51–57.
Wolff, L. E., and Duncan, J. L. (1974). *J. gen. Microbiol.*, **81,** 413–424.
Yamamoto, T., Imai, S., Nisizawa, T., and Araya, S. (1975). *Archs. oral Biol.*, **20,** 389–391.

CHAPTER XI

# Laboratory Diagnosis, Serology and Epidemiology of *Streptococcus pneumoniae*

ERNA LUND

*The Pneumococcus Department, Statens Seruminstitut,*
*Copenhagen, Denmark*

AND

JØRGEN HENRICHSEN

*The Streptococcus Department, Statens Seruminstitut,*
*Copenhagen, Denmark*

# I. INTRODUCTION

*Streptococcus pneumoniae* (= pneumococcus) is a species of the genus *Streptococcus* of the family *Streptococcaceae* (Buchanan and Gibbons, 1974).

Pneumococci are Gram-positive, capsulated, lanceolate diplococci often arranged in short, straight chains; typically, they grow diffusely in serum broth and give rise to smooth colonies surrounded by narrow zones of α-haemolysis on blood agar; they are facultative anaerobes and some strains are carbon dioxide dependent upon isolation from clinical specimens; they are killed by heat at 60°C for 30 min, easily lysed, soluble in bile, sensitive to optochin and many types are virulent for mice.

# II. THE DIAGNOSIS OF *STREPTOCOCCUS PNEUMONIAE*

Fluid specimens and swabs are inoculated onto blood agar and into serum broth. The cultures are incubated at 37°C. On blood agar, typical pneumococcal colonies may be observed as round, flat, smooth, translucent, often with a central pitting and with a greenish discoloration of the surrounding medium. Rarely, non-capsulated strains with rough (R) form colonies are isolated. Among the smooth (S) form colonies of capsulated strains, non-capsulated, rough (R) form mutant colonies are often seen to emerge spontaneously in laboratory cultures.

If an isolated strain of Gram-positive diplococci grows with typical colonies on blood agar and diffusely in serum broth, the final diagnosis of *S. pneumoniae* becomes a matter of differentiating between this and other species of α-haemolytic streptococci. This may be done by examining the strain in question for its sensitivity to optochin and its capsular reaction in diagnostic pneumococcal sera.

## A. Optochin test (Lund, 1959)

Sensitivity to optochin (ethylhydrocupreine hydrochloride) may be examined by inoculating a young culture on blood agar containing 0·05% (w/v) optochin and, as a control, also on blood agar without optochin. Pneumococci are not able to grow on this optochin blood agar, while all other streptococci grow well. Alternatively this test may also conveniently be performed by means of tablets or paper discs containing optochin.

Tablets may be obtained commercially (Diatabs[R] produced by Rosco Ltd., DK-2630 Taastrup, Denmark). Discs can be produced by moistening circular filter paper discs (with a diameter of 6 mm) in a 0·05% (w/v) solution of optochin in distilled water and thereafter drying the discs in air. The test is carried out by placing a tablet or a disc on a rather densely inoculated blood agar plate and measuring the diameter of the growth inhibition zone after aerobic incubation. All strains of S and R pneumococci examined have inhibition zones with diameters of 20 mm or more, while other streptococci have zones of less than 12 mm, a clear-cut difference in sensitivity to optochin which makes this reagent highly suitable for differentiation.

## B. Capsular reaction test (Neufeld test) (Neufeld, 1902)

In most instances it suffices to know whether a given strain is a pneumococcus or not. If the strain is sensitive to optochin, the diagnosis may be confirmed using the "Omni serum" which is a pooled polyvalent, concentrated pneumococcal serum giving a capsular reaction in a Neufeld test with all 83 known types of pneumococci (Lund and Rasmussen, 1966; Lund *et al.*, 1972).

Examination for capsular reaction may be carried out as follows: A tiny amount of a broth culture (or a suspension of bacteria) is placed on a slide. A little serum is mixed thoroughly with the droplet on the slide. It is preferable to have relatively few organisms per microscope field. A coverslip is placed over the mixture and the preparation is examined under the microscope with an oil immersion lens for capsular reaction and for agglutination of the bacteria. The Neufeld reaction is not a swelling of the capsule but a reaction between the type-specific serum and the capsular substance rendering the capsule visible. It is also called the quellung reaction and has recently been described in detail by Austrian (1976).

Not all the type-specific antibodies of Omni serum, however, are present in titres high enough to allow capsules to become visible, but produce agglutination, which must sometimes be accepted as the sole criterion of a positive reaction.

To establish a type diagnosis of a pneumococcus, the strain is first examined in the pooled sera and thereafter in the relevant type or group sera. R forms may give non-specific agglutination in the sera of several types.

The diagnostic pneumococcal sera, of course, may also be used for direct examination of clinical specimens such as sputa, pleural exudates and spinal fluids, either by microscopy or by counter-current immunoelectrophoresis (Colding and Lind, 1977; Coonrod and Rytel, 1972; El-Refaie and Dulake, 1975). Because they are neutral, the counter-current immunoelectrophoresis technique fails to demonstrate the presence of the capsular polysaccharide antigens of the frequently disease-causing types 7F and 14, unless a special—and complicated—buffer system is used (Anhalt and Yu, 1975).

## III. CAPSULAR ANTIGENS OF PNEUMOCOCCI

The capsules of pneumococci are made up of polysaccharides varying in their mono- and disaccharide components. Table I lists the presently known 83 different pneumococcal serotypes together with their antigenic formulae. Some of the serotypes form serogroups, viz. types carrying the same number, but different capital letters, e.g. serotypes 7F, 7A, 7B and 7C form serogroup 7. The quite extensive serological cross-reactivity between some of the different types indicates the possession of common antigenic determinants, probably best explained by the existence of similar or identical mono-, di-, or oligo-saccharide units in the capsular polysaccharides of the different serotypes. The table is not to be regarded as complete; new types might still be discovered and involve the addition of new antigens or antigenic determinants. Some of the types have no or very few cross-reactions, e.g. type 1 = 1a, type 2 = 2a, while other types have antigens in common with several types, e.g. types 7B, 7C, 19B, 19C, 24F, 24B and 40 have the antigen 7h in common.

As early as 1928 it was shown by Griffith that the serotype of a pneumococcus may be changed by transformation in *in vivo* experiments. Since then much work has been carried out on the transformation of pneumococci (see Hayes, 1968, for review), but it is still not known with certainty whether transformation occurs spontaneously *in vivo*.

The pneumococcal capsular polysaccharide has been called "specific soluble substance" (SSS). It may be prepared by the following procedure from strains of pneumococci (Roesgaard, 1945): The strain is inoculated into 100 ml of a 5% serum broth which is incubated at 37°C for 18 h; 5–10 ml of ox bile is then added. After 2 h at 37°C, the culture will be autolysed. One millilitre of 50% acetic acid is added and the mixture heated to boiling point. After cooling, it is neutralised with sodium hydroxide and centrifuged for 20 min at 4500 rpm. The type-specific polysaccharide is precipitated from the supernatant with two volumes of absolute alcohol, left at 4°C overnight, and then sedimented by centrifugation for 10 min at 3000 rpm, washed with 10–20 ml absolute alcohol, centrifuged again and dried in a desiccator for 10–12 h. Drying may be completed by shaking with ether and leaving in the incubator. The dried sediment is ground to a fine powder and is stable to prolonged storage at refrigerator temperatures.

So far, precipitation of SSS from the urine of patients with pneumococcal infections has had little practical diagnostic value, but might prove to be of use in connection with sensitive immunochemical methods, e.g. counter-current immunoelectrophoresis.

For the demonstration of free capsular polysaccharide in serum, radioimmunoassay seems to be the method of choice (Schiffman *et al.*, 1974).

# TABLE I

Danish type designations and antigenic formulae of 83
pneumococcal serotypes (Kauffmann and Lund, 1954;
Lund, 1970a)

| Type | Antigenic formula | Type | Antigenic formula |
|------|-------------------|------|-------------------|
| 1    | 1a                | 20   | 20a, 20b, 7g      |
| 2    | 2a                | 21   | 21a               |
| 3    | 3a                | 22F  | 22a, 22b          |
| 4    | 4a                | 22A  | 22a, 22c          |
| 5    | 5a                | 23F  | 23a, 23b, 18b     |
| 6A   | 6a, 6b            | 23A  | 23a, 23c, 15c     |
| 6B   | 6a, 6c            | 23B  | 23a, 23b, 23d     |
| 7F   | 7a, 7b            | 24F  | 24a, 24b, 24d, 7h |
| 7A   | 7a, 7b, 7c        | 24A  | 24a, 24c, 24d     |
| 7B   | 7a, 7d, 7e, 7h    | 24B  | 24a, 24b, 24e, 7h |
| 7C   | 7a, 7d, 7f, 7g, 7h| 25   | 25a, 25b          |
| 8    | 8a                | 27   | 27a, 27b          |
| 9A   | 9a, 9c, 9d        | 28F  | 28a, 28b, 16b, 23d|
| 9L   | 9a, 9b, 9c, 9f    | 28A  | 28a, 28c, 23d     |
| 9N   | 9a, 9b, 9e        | 29   | 29a, 29b, 13b     |
| 9V   | 9a, 9c, 9d, 9g    | 31   | 31a, 20b          |
| 10F  | 10a, 10b          | 32F  | 32a, 27b          |
| 10A  | 10a, 10c, 10d     | 32A  | 32a, 32b, 27b     |
| 11F  | 11a, 11b, 11c, 11g| 33F  | 33a, 33b, 33d     |
| 11A  | 11a, 11c, 11d, 11e| 33A  | 33a, 33b, 33d, 20b|
| 11B  | 11a, 11b, 11f, 11g| 33B  | 33a, 33c, 33d, 33f|
| 11C  | 11a, 11b, 11c, 11d, 11f | 33C | 33a, 33c, 33e |
| 12F  | 12a, 12b, 12d     | 34   | 34a, 34b          |
| 12A  | 12a, 12c, 12d     | 35F  | 35a, 35b, 34b     |
| 13   | 13a, 13b          | 35A  | 35a, 35c, 20b     |
| 14   | 14a               | 35B  | 35a, 35c, 29b     |
| 15F  | 15a, 15b, 15c, 15f| 35C  | 35a, 35c, 20b, 42a|
| 15A  | 15a, 15c, 15d, 15g| 36   | 36a, 9e           |
| 15B  | 15a, 15b, 15d, 15e, 15h | 37 | 37a            |
| 15C  | 15a, 15d, 15e     | 38   | 38a, 25b          |
| 16   | 16a, 16b, 11d     | 39   | 39a, 10d          |
| 17F  | 17a, 17b          | 40   | 40a, 7g, 7h       |
| 17A  | 17a, 17c          | 41F  | 41a, 41b          |
| 18F  | 18a, 18b, 18c, 18f| 41A  | 41a               |
| 18A  | 18a, 18b, 18d     | 42   | 42a, 20b, 35c     |
| 18B  | 18a, 18b, 18e, 18g| 43   | 43a, 43b          |
| 18C  | 18a, 18b, 18c, 18e| 44   | 44a, 44b, 12b, 12d|
| 19F  | 19a, 19b, 19d     | 45   | 45a               |
| 19A  | 19a, 19c, 19d     | 46   | 46a, 12c, 44b     |
| 19B  | 19a, 19c, 19e, 7h | 47F  | 47a, 35a, 35b     |
| 19C  | 19a, 19c, 19f, 7h | 47A  | 47a, 43b          |
|      |                   | 48   | 48a               |

## IV. SEROLOGICAL CROSS-REACTIONS BETWEEN POLYSACCHARIDES OF PNEUMOCOCCI AND OF OTHER BACTERIA

In 1944, the isolation of a capsulated non-haemolytic streptococcus cross-reacting with pneumococcal types 19A, 19B, 19C and 40 was reported (Mørch, 1944). This finding was later extended by Lund (1950) and the whole question of cross-reactivity between pneumococcal and streptococcal polysaccharides has recently been reviewed by Austrian (1973).

There are also several examples of serological cross-reactions between pneumococci and Gram-negative bacteria. Thus *Escherichia coli* antigens K30, K42, and K85 were shown to precipitate pneumococcal antisera of types 2 and 5; 1 and 25; and 2, 5, 10 and 23, respectively (Heidelberger *et al.*, 1968) and pneumococci of group 9 and of type 27 have antigenic determinants in common with strains of *E. coli*, type 2 with *Klebsiella* (Mørch and Knipschildt, 1944) and group 35 with *Salmonella kirkee* (Kauffman and Langvad-Nielsen, 1942). Several pneumococcal types cross-react with *Haemophilus influenzae* (Engbaek, 1949).

## V. PRODUCTION OF DIAGNOSTIC PNEUMOCOCCAL SERA

### A. Vaccine

The strain employed for the preparation of vaccine for rabbit or human immunisation must have well developed capsules which give a clear and indisputable capsular reaction with the homologous serum. As a rule 15–18 litres of trypsin broth† are inoculated for each strain, using five or six flasks of 3 litres. Before inoculation these flasks are left in the incubator at 37°C for 24 h as a sterility check. In the morning each 3-litre flask is inoculated with 20 ml of an overnight serum broth culture. The flasks are shaken and incubated again at 37°C. Growth is followed by removing samples of 5–10 ml from time to time, estimating their density and testing the capsule formation by performing a Neufeld test with the homologous serum. Particular attention must be paid to cultures of types 3 and 37 and of group 19, as the capsules of these types degenerate very rapidly; it may sometimes be necessary to interrupt growth before it has reached its maximum. In most cases, however, the capsules are well preserved when the maximum density of the culture has been reached after 4–7 h growth.

Growth is stopped by the addition of formalin to a concentration of 2%,

† The trypsin both is prepared essentially as described by Pope and Smith (1932) the only modification being that the trypsin digestion is discontinued when an analysis has shown the aminonitrogen content to be at least 1 mg per ml.

after subcultures from each flask have been made onto blood agar plates as a control of purity. The flasks are left at room temperature until the following morning, when the cultures are centrifuged. The supernatant is discarded and the sediment is ground in a mortar and resuspended in Sørensen buffer solution (20 ml of $M/15$ $KH_2PO_4$ + 80 ml of $M/15$ $Na_2HPO_4$, $2H_2O$ + 300 ml of $0.9\%$ NaCl solution) with $0.5\%$ formalin. The density of the vaccine is estimated by means of standards containing 1000–2000, 2000–4000 and 4000–8000 million pneumococci per ml and designated 1, 2 and 4, respectively. Vaccines are stored as suspensions that are 15–25 times denser than standard 1.

Capsular reactions of the vaccines are checked before use with the homologous serum and a Gram stain is also performed. In a successful vaccine all the bacteria are Gram-positive. When stored in dense suspensions at $4°C$, the vaccines of most of the pneumococcal types have been found to preserve their capacity for capsular reaction and agglutination for several years. Exceptions to this rule, however, are the vaccines of types 3 and 37 and of group 19, which must be replaced after 2–3 months.

## B. Immunisation

White rabbits weighing 2–3 kg are used for immunisation. The vaccine is heated to $37°C$ before it is injected slowly into an ear vein. The usual procedure is to give the rabbits three injections a week (i.e. every other day) and vary the dosage as follows: 1st week: $1.0$ ml of standard 2, 2nd to 4th week: $1.0$ ml of standard 4. After 4 or 5 weeks the animals are bled by heart puncture. About 50 ml blood is taken from each rabbit. Six to seven days after heart puncture, the immunisation is resumed, and 3–4 weeks later heart puncture is repeated. When the immunisation period is more protracted, the number of injections of vaccine may be reduced to one or two a week. At the conclusion of the immunisation the animals are bled out.

It is possible by immunising rabbits with a vaccine mixture to produce sera that react with several types. A serum that is to be used, e.g., for diagnosis of group 7 (types 7F, 7A, 7B, 7C) may be obtained by immunisation with a vaccine containing equal parts of vaccine of types 7F, 7A, 7B and 7C. The dosage schedule for such an immunisation may be the same as that given for a vaccine of one type. If necessary, higher doses may be tried, possibly with a denser vaccine.

Capsular and agglutination titres of each serum obtained by heart puncture are determined. Sera from several heart punctures of equal strength may be pooled.

Titration is carried out in small tubes containing $0.2$ ml of doubling dilutions: 1:1, 1:2, 1:4, etc. A control test for auto-agglutination of the vaccine is made in a tube with $0.2$ ml physiological saline. To each tube is added 0·2 ml of pneumococcal vaccine of a standardised density (about 100 million cells per ml). The tubes are shaken and a loopful of the contents of each is examined under the microscope. The reciprocal value of the highest dilution of serum still giving an unquestionable agglutination, but not necessarily a visible quellung reaction, indicates the "capsular titre".

A diagnostic serum must be specific, i.e. it must react only with the type or group against which it has been raised. Each portion of serum must therefore be examined for cross-reactions. This is carried out by means of Neufeld tests with vaccines of all known pneumococcal types as used for determining capsular and agglutination titres. The degree of the cross-reactions found is assessed by titration, and these reactions eliminated by absorption.

## C. Absorption

Absorption is carried out with the dense formolised vaccines. It is possible with some experience to estimate the amount of vaccine required to remove a cross-reaction of a certain degree. The vaccine is centrifuged for 30–60 min at a rate of 500 rpm and the clear supernatant fluid discarded. The serum to be absorbed is poured into the centrifuge tube and mixed thoroughly with the vaccine sediment by means of a Pasteur pipette. The absorption is completed within 5–10 min, irrespective of temperature and serum concentration.

The mixture of serum and vaccine is centrifuged for 30–60 min at the same rate as before. The clear supernatant serum is pipetted off and tested for capsular reaction with the absorbing type to ensure that the absorption has left no residual cross-reactivity. A cross-reaction with a capsular titre of 32–64 can usually be removed by one absorption, but with higher titres repeated absorption may be necessary. A control test is thereafter again made on the serum. If several reactions with low titres (2–8) have to be removed from the same serum, two or three different vaccines may be used successfully for simultaneous absorption.

After filtration the final capsular titres are determined. It is general experience that absorbed sera decrease in potency more rapidly than non-absorbed sera, and hence it is necessary at intervals of a few months to check the capsular titres of the diagnostic sera, nearly all of which are absorbed.

Five kinds of pneumococcal sera are employed for diagnostic purposes:

(a)  omni serum, reacting with all 83 types,
(b)  type sera, reacting with a single type,
(c)  group sera, reacting with all the types within one group,

(d)  pooled sera, reacting with several different types or groups,
(e)  factor sera, absorbed sera employed for differential diagnosis within a group.

Statens Seruminstitut in Copenhagen produces the following diagnostic pneumococcal sera for sale:

1. *Omni serum*
2. *Nine pooled sera, A–I*, each reacting with 7–11 types, covering all 83 types
3. 46 *type or group sera* which react with single types or groups, numbered 1–48. Nos. 26 and 30 (now types 6B and 15A) are not in use.

## D. Omni serum

An Omni serum is a pool of equal amounts of the nine (A–I) polyvalent sera. The pooled serum is purified and concentrated by salting out with sodium sulphate. The precipitate is dialysed, lyophilised and re-dissolved in saline to a quarter of the original volume (Lund and Rasmussen, 1966). The Omni serum gives a capsular reaction with all 83 known types. It is primarily intended for use as a quick diagnostic tool directly on clinical specimens, especially spinal fluids.

## E. Type sera

By immunisation with individual serotype strains type sera are produced. Heterologous reactions are absorbed out. These sera are used for the diagnosis of types that do not form serogroups, while absorbed factor sera (*vide infra*) containing antibodies for the specific antigens of each type are used for differentiation of types within a group.

It is impossible to produce type-specific sera for a few types, since these contain no antigen specific for the type but consist exclusively of antigens that are present also in other pneumococcal types. When, for instance, a serum 7F (7a, 7b) is absorbed with organisms of type 7A (7a, 7b, 7c), the serum is depleted of all antibodies and gives no reaction at all. The same applies to sera of types 32F, 33F and 41A. In such cases we use the stronger group sera (*vide infra*) obtained by immunisation with related types.

The Danish diagnostic pneumococcal type sera 29 and 42 are not type-specific. If all heterologous reactions were absorbed from these sera, the homologous titres would be too low. Type sera 29 and 42 have the following heterologous reactions:

Serum Pn 29 reacts with type 35B
Serum Pn 42 reacts with types 20, 31 and groups 33, 35

## F. Group sera

Polyvalent group sera may be obtained by pooling the monovalent sera corresponding to the individual types making up the group. As it is also possible to produce these sera by immunisation with several types injected simultaneously, it is customary to do so in order to avoid the dilution brought about by the pooling procedure. The immune serum is freed from heterologous reactions by absorption and the capsular titre determined for each type of the group.

For the diagnosis of group 35 it is preferable to use a serum produced with type 35F alone because such a serum gives good reactions with all the types of the group (35F, 35A, 35B, 35C), while group 35 serum produced by immunisation with all the types in the group gives such strong heterologous reactions that the removal of these leaves only a rather weak diagnostic serum. Also the group 35 serum-like type sera 29 and 42 — is not specific as it also reacts with types 42 and 47F.

## G. Pooled sera A–I

These nine pooled sera are composed in such a way that each of the 83 types known gives a capsular reaction in one of these sera. The distribution is planned with a view to the cross-reactions: the types which have strong reactions with each other are gathered in the same pool, thereby rendering the removal of these reactions unnecessary.

These pooled sera may be produced by immunisation with up to 11 different types simultaneously, although the use of more than five strains at a time leads to a reduction in the amount of type-specific antibody formed. The pooled sera A–I consist of the following types or groups:

| Pool | Types or groups | | | | | | |
|------|----|----|----|----|----|----|----|
| A | 1  | 2  | 4  | 5  | 18 | | |
| B | 3  | 6  | 8  | 19 | | | |
| C | 7  | 20 | 24 | 31 | 40 | | |
| D | 9  | 11 | 16 | 36 | 37 | | |
| E | 10 | 12 | 21 | 33 | 39 | | |
| F | 17 | 22 | 27 | 32 | 41 | | |
| G | 29 | 34 | 35 | 42 | 47 | | |
| H | 13 | 14 | 15 | 23 | 28 | | |
| I | 25 | 38 | 43 | 44 | 45 | 46 | 48 |

## H. Factor sera

Factor sera (Table II) are especially absorbed sera used for differential diagnosis within a pneumococcal group. For instance, in order to distinguish between the two types in group 6, 6A and 6B, two sera are required, one of which reacts only with type 6A, the other only with type 6B. On absorption of serum 6A with type 6B, the serum components that react with 6B are removed, and the absorbed serum will then give a reaction only with type 6A. Correspondingly, the serum specific for type B is obtained by absorption of serum 6B with type 6A. The two types are set up with the following formulae for their capsular antigenic determinants:

6A = 6a, 6b
6B = 6a, 6c

6a signifies the antigen (or antigens) common to the two types, while 6b and 6c signify the antigen specific for each type. By absorption of serum 6A with type 6B, the mutual component 6a is removed, leaving a serum 6b that is specific for type 6A. Absorption of serum 6B with type 6A removes the factor 6a too, leaving serum 6c that reacts only with type 6B. By means of these two sera, 6b and 6c, it is possible to make the differential diagnosis within group 6.

In the case of a group consisting of more than two types, the procedure is fundamentally the same. From the antigenic formulae it is possible to determine the absorption processes required.

These strongly absorbed factor sera weaken rather rapidly. They are not commercially available.

### I. Stability of the sera

Diagnostic pneumococcal sera are stable for 1–2 years or more if kept at 4°C. Stability depends on the titres and on the degree of absorption. Non-absorbed serum has proved to be useful after storage for 10–20 or even 30 years (at 4°C). Thiomersalate is added as a preservative to a concentration of 0·01%.

### J. Titres of the sera

The National Institutes of Health (USA) required the following titres for diagnostic pneumococcal sera:

Type sera: 16, pooled sera: 8, type 3 serum: 8

The Danish sera follow the same rules. The group sera have a titre of at least 16 for each type and the polyvalent Omni serum, covering all 83 types, has a minimum titre of 4.

### K. The importance of typing strains of *S. pneumoniae*

Although *S. pneumoniae* can easily be identified without use of pneumococcal sera, omni serum provides a means of achieving a species diagnosis

TABLE II
## Pneumococcal factor sera

| Type | Antigenic formula | Factor sera | | | | |
|------|-------------------|------|------|------|------|------|
| | | 6b | 6c | | | |
| 6A | 6a, 6b | + | − | | | |
| 6B | 6a, 6c | − | + | | | |
| | | 7b | 7c | 7e | 7f | |
| 7F | 7a, 7b | + | − | − | − | |
| 7A | 7a, 7b, 7c | + | + | − | − | |
| 7B | 7a, 7d, 7e, 7h | − | − | + | − | |
| 7C | 7a, 7d, 7f, 7g, 7h | − | − | − | + | |
| | | 9b | 9c | 9d | 9e | 9g |
| 9A | 9a, 9c, 9d | − | + | + | − | − |
| 9L | 9a, 9b, 9c, 9f | + | (+) | − | − | − |
| 9N | 9a, 9b, 9e | + | − | − | + | − |
| 9V | 9a, 9c, 9d, 9g | − | (+) | + | − | + |
| | | 10b | 10c | | | |
| 10F | 10a, 10b | + | − | | | |
| 10A | 10a, 10c | − | + | | | |
| | | 11b | 11c | 11f | | |
| 11F | 11a, 11b, 11e, 11g | + | − | − | | |
| 11A | 11a, 11c, 11d, 11e | − | + | − | | |
| 11B | 11a, 11b, 11f, 11g | + | − | + | | |
| 11C | 11a, 11b, 11c, 11d, 11f | − | (+) | + | | |
| | | 12b | 12c | | | |
| 12F | 12a, 12b, 12d | + | − | | | |
| 12A | 12a, 12c, 12d | − | + | | | |
| | | 15b | 15c | 15d | 15e | 15h |
| 15F | 15a, 15b, 15c, 15f | + | + | − | − | − |
| 15A | 15a, 15c, 15d, 15g | − | (+) | (+) | − | − |
| 15B | 15a, 15b, 15d, 15e, 15h | + | − | (+) | + | + |
| 15C | 15a, 15d, 15e | − | − | + | + | − |
| | | 17b | 17c | | | |
| 17F | 17a, 17b | + | − | | | |
| 17A | 17a, 77c | − | + | | | |

Table II (*continued*)

| Type | Antigenic formula | Factor sera | | | |
|------|-------------------|------|------|------|------|
| | | 18c | 18d | 18e | 18f |
| 18F | 18a, 18b, 18c, 18f | + | − | − | + |
| 18A | 18a, 18b, 18d | − | + | − | − |
| 18B | 18a, 18b, 18e, 18g | − | − | + | − |
| 18C | 18a, 18b, 18c, 18e | + | − | (+) | − |
| | | 19b | 19c | 7h | 19f |
| 19F | 19a, 19b, 19d | + | − | − | − |
| 19A | 19a, 19c, 19d | − | + | − | − |
| 19B | 19a, 19c, 19e, 7h | − | − | + | − |
| 19C | 19a, 19c, 19f, 7h | − | − | + | + |
| | | 22b | 22c | | |
| 22F | 22a, 22b | + | − | | |
| 22A | 22a, 22c | − | + | | |
| | | 23b | 23c | 28 | |
| 23F | 23a, 23b, 18b | + | − | − | |
| 23A | 23a, 23c, 15a | − | + | − | |
| 23B | 23a, 23b, 23d | (+) | − | + | |
| | | 24c | 24d | 24e | |
| 24F | 24a, 24b, 24d, 7h | − | + | − | |
| 24A | 24a, 24c, 24d | + | + | − | |
| 24B | 24a, 24b, 24e, 7h | − | − | + | |
| | | 28b | 28c | | |
| 28F | 28a, 28b, 16b, 23d | + | − | | |
| 28A | 28a, 28c, 23d | − | + | | |
| | | 32a | 32b | | |
| 32F | 32a, 27b | + | − | | |
| 32A | 32a, 32b, 27b | + | + | | |
| | | 33b | 20b | 33f | 33e |
| 33F | 33a, 33b, 33d | + | − | − | − |
| 33A | 33a, 33b, 33d, 20b | + | + | − | − |
| 33B | 33a, 33c, 33d, 33f | − | − | + | − |
| 33C | 33a, 33c, 33e | − | − | − | + |

Table II (*continued*)

| Type | Antigenic formula | Factor sera | | | |
|------|-------------------|------|------|------|------|
| | | 35b | 35c | 29b | 42a |
| 35F | 35a, 35b, 34b | + | − | − | − |
| 35A | 35a, 35c, 20b | − | + | − | − |
| 35B | 35a, 35c, 29b | − | (+) | + | − |
| 35C | 35a, 35c, 20b, 42a | − | + | − | + |
| | | 41a | 41b | | |
| 41F | 41a, 41b | + | + | | |
| 41A | 41a | + | − | | |
| | | 47a | 43b | | |
| 47F | 47a, 35a, 35b | + | − | | |
| 47A | 47a, 43b | + | + | | |

+, distinct capsular reaction, (+), weak capsular reaction, −, no capsular reaction.

sooner. As serum therapy is no longer in use typing of pneumococci has no bearing on the treatment of the individual, acutely ill patient. However, because the prevalence of different types varies with time and geographical area, typing surveillance studies must be performed throughout the world to ensure optimal composition of pneumococcal polysaccharide vaccines (*vide infra*) for use in different areas, as only a limited number of types can be included. The currently available 14-valent vaccine is made up of types mainly chosen on the basis of studies performed during recent years in the U.S.A. Probably, the 10–20 most commonly occurring types will be found to be roughly the same all over the world, even though their relative position might vary. Also typing of disease isolates from vaccinees must be continued to see whether vaccine usage leads to the predominance of different types; so far, this does not seem to be the case (Austrian *et al.*, 1976).

## VI. NOMENCLATURE OF THE PNEUMOCOCCAL TYPES

In 1960, a brief history was given of the nomenclature of pneumococcal types and a correlation made of the Danish and American serotype designations (Kauffmann *et al.*, 1960). Since then, three new types have been described: Pn 48 (Lund, 1962), 12A (Lund and Munksgaard 1967) and 47A (Lund *et al.*, 1972). For some years now, it has been possible at Statens Seruminstitut, Copenhagen, to type all capsulated pneumococci received from patients in Denmark (Lund, 1970b) and other countries.

The Danish nomenclature was first published by Kaufmann *et al.* (1940) and later extended by Lund (Kauffmann and Lund, 1954; Lund, 1970a). The American nomenclature was established by Eddy (1944) who gave the various types consecutive numbers (1–81) regardless of their antigenic structure. The Kauffmann-Lund nomenclature is based on antigenic relationships: Types that possess common capsular antigens compose natural groups. Forty-six type or group sera (Pn 1–48, Nos. 26 and 30 excluded), nine pooled sera (A–I) and one Omni serum cover all 83 types.

In a publication on pneumococcal types, Lund (1970a) gives formulae for 82 types. However, there are certain difficulties involved in this system. The first established type in a group is designated by its original name, being a number without any letter added, e.g. type 7, while the other types in the group are called both by the number and a letter: 7A, 7B and 7C. Thus, when the result of a typing is given as Pn 7, it is not clear whether it is the original type 7 (antigenic formula 7a, 7b) or whether it is the diagnosis of the group, indicating that this strain belongs to group 7, but that further typing within the group has not been carried out.

As serological typing is not subject to formally recognised international recommendations, we propose to add an F (the first type) to the first established types in 16 groups, so that e.g. the 7 group comprises types 7F, 7A, 7B and 7C, and the 10 group types 10F and 10A (Table II). It is proposed to leave the nomenclature of the 6 and 9 groups, where the original convention has not been followed, unchanged and only add the letter F in the other groups.

This altered nomenclature is recommended as the international system.


## VII. EPIDEMIOLOGY OF *S. PNEUMONIAE*

Pneumococci may cause infections of the middle ear, the sinuses and the lungs (lobar pneumonia) and are also frequently isolated from cases of purulent meningitis. Recently, *S. pneumoniae* was reported to be the single most common causative organism of bacteraemia at Boston City Hospital (McGowan *et al.*, 1975). More rarely, pneumococci are isolated from arthritis, pericarditis, eye infections, abscesses, and skin infections. In 1933, pneumococci were demonstrated in approximately 50% of throat cultures from healthy people (Gundel, 1933).

In Table III, the most commonly occurring pneumococcal types or groups in Denmark in the years 1939–1947 and 1955–1970 are compared. In the first period type 1 was the type most frequently isolated from blood, spinal fluid, pleural exudate and pus from ears and the second most

TABLE III

**Comparison of the most common pneumococcal types in 1939-1947 and 1955-1970, listed in order of frequency**

| Material | 1939–1947 | | | | | | | 1955–1970 | | | | | | |
|---|---|---|---|---|---|---|---|---|---|---|---|---|---|---|
| Blood | 1 | 2 | 3 | 6 | 4 | 12 | 23 | 14 | 4 | 1 | 7 | 3 | 23 | 9 |
| Spinal fluid | 1 | 3 | 6 | 18 | 2 | 4 | 19 | 18 | 14 | 7 | 6 | 3 | 23 | 19 |
| Pleural exudate | 1 | 2 | 6 | 3 | 19 | 7 | 8 | 3 | 23 | 1 | 14 | 19 | 9 | 7 |
| Expectoration | 3 | 1 | 2 | 7 | 6 | 19 | 8 | 6 | 19 | 3 | 23 | 9 | 17 | 15 |
| Swabs | 6 | 19 | 1 | 14 | 3 | 18 | 23 | 19 | 6 | 3 | 23 | 14 | 9 | 11 |
| Ear | 1 | 3 | 19 | 6 | 14 | 5 | 18 | 19 | 14 | 6 | 23 | 3 | 18 | 1 |
| Sinus maxillaris* | 6 | 3 | 19 | 23 | 22 | | | 6 | 23 | 19 | 8 | 34 | 18 | 7 |

\* The figures are from 1941–1947.

## TABLE IV

**Distribution of pneumococcal types isolated in Denmark from various clinical specimens during the years 1955–1970 (2684 strains). Types found in < 1% are omitted**

| Type/ group | Blood | Spinal fluid | Pleural exudate | Expect- oration | Swab | Ear | Nose | Total Number | % | Fre- quency |
|---|---|---|---|---|---|---|---|---|---|---|
| 1 | 44 | 14 | 15 | 4 | 4 | 19 | 3 | 103 | 3·8 | 10 |
| 2 | 8 | 20 | 7 | 6 | 2 | 1 | — | 44 | 1·6 | 19 |
| 3 | 37 | 46 | 18 | 34 | 24 | 32 | 9 | 200 | 7·4 | 5 |
| 4 | 45 | 31 | 3 | 5 | 3 | 2 | 2 | 91 | 3·4 | 11 |
| 6 | 29 | 52 | 8 | 41 | 42 | 35 | 38 | 245 | 9·1 | 2 |
| 7 | 38 | 53 | 9 | 12 | 12 | 12 | 10 | 146 | 5·4 | 6 |
| 8 | 29 | 42 | 8 | 12 | 8 | 15 | 14 | 128 | 4·8 | 8 |
| 9 | 30 | 28 | 10 | 27 | 15 | 8 | 10 | 128 | 4·8 | 9 |
| 10 | 5 | 9 | 6 | 9 | 7 | 6 | 3 | 45 | 1·7 | 18 |
| 11 | 7 | 8 | 2 | 11 | 15 | 1 | 10 | 54 | 2·0 | 16 |
| 12 | 23 | 21 | 1 | 3 | 2 | 4 | 2 | 56 | 2·1 | 14 |
| 14 | 50 | 56 | 15 | 14 | 16 | 50 | 7 | 208 | 7·7 | 3 |
| 15 | 10 | 13 | 6 | 18 | 14 | 11 | 8 | 80 | 2·9 | 12 |
| 16 | 3 | 6 | 3 | 8 | 3 | 4 | 6 | 33 | 1·2 | 21 |
| 17 | 9 | 4 | 2 | 21 | 5 | 6 | 8 | 55 | 2·0 | 15 |
| 18 | 19 | 61 | 5 | 7 | 15 | 21 | 11 | 139 | 5·1 | 7 |
| 19 | 26 | 44 | 14 | 36 | 51 | 94 | 27 | 292 | 10·8 | 1 |
| 20 | 7 | 5 | 4 | 5 | 3 | 1 | 6 | 31 | 1·2 | 22 |
| 22 | 14 | 9 | 8 | 14 | 5 | 1 | 1 | 52 | 1·9 | 17 |
| 23 | 32 | 45 | 16 | 29 | 22 | 34 | 28 | 206 | 7·6 | 4 |
| 31 | 3 | 2 | 5 | 10 | 6 | — | 1 | 27 | 1·0 | 24 |
| 33 | 8 | 10 | 2 | 5 | 2 | 1 | 1 | 29 | 1·1 | 23 |
| 34 | 4 | 14 | 6 | 18 | 12 | — | 12 | 66 | 2·4 | 13 |
| 35 | 2 | 4 | 5 | 7 | 10 | — | 7 | 35 | 1·3 | 20 |

frequently isolated from sputum. Types 2 and 6 were also among the dominant types. During the second period, the pattern was different. Type 1 was not the most frequent in any material and was not even among the seven most frequently isolated types from spinal fluid and sputum; and type 2 had fallen to only a very low incidence.

Table IV gives the type distribution of 2684 strains isolated in Denmark from various clinical specimens during the years 1955–1970. The types found with a frequency of less than 1% are omitted. The seven most frequent types or groups were 19, 6, 14, 23, 3, 7 and 18. Type 1 was number 10 and type 2 number 19. Type 14 was isolated most frequently from blood, group 18 from spinal fluid, type 3 from pleural exudate, group 6 from sputum and pus from nose and sinus maxillaris. In swabs from throat, larynx and ears, group 19 was dominant. Group 19 was most frequent overall.

Table V (1962–1969) shows the frequency of the types within the groups. It will be seen from this table that types 6A and 6B were found almost equally often, while type 7F was by far the most frequent type in group 7 (7F, 7A, 7B, 7C). In the 9 group (9A, 9L, 9N, 9V), type 9N predominated, in the 18 group (18F, 18A, 18B, 18C) type 18C, and in the 23 group (23F, 23A, 23B) type 23F. In the 19 group (19F, 19A, 19B, 19C) types 19F and 19A were isolated quite frequently, while types 19B and 19C were not found at all in the period 1962–1969.

Lobar pneumonia is found more often in men than in women (Bullowa, 1937; Mørch-Lund, 1949) and especially in young and middle-aged people. In Denmark, it is relatively infrequent approximating only 5000 cases per year — in a population of more than 5 million people.

Different types of pneumococci have been isolated from cases of lobar pneumonia at different times in different areas. This is exemplified in Table VI which shows the distribution of types 1 and 2 over the years in different countries. In the USA, type 1 was, in 1913, found in 47% and type 2 in 18% of cases. The following year in the same hospital, type 1 was seen in 30% and type 2 in 39%, the change for type 2 in one year thus being from 18% to 39%. Other authors have reported similar figures, showing that type 2 varied greatly, while type 1 was relatively constant.

In South Africa, Lister found in 1913–1917 that type 5 was the most frequent type of pneumonia, namely 31%, while type 1 was seen in 22% and type 2 in 16% of cases. Ordman continued these investigations and in 1938 reported type 1 in 22% and type 2 in 15%. The strong representation of type 5 had disappeared. Among native miners type 2 dominated at that time.

In Denmark in 1923, Christensen found type 1 in 33% of cases of lobar pneumonia, and type 2 in 27%. Comparing these findings with the results of Nissen in 1937, it was seen that type 2 had decreased from 27% to 1·7%, type 1 being the most frequent in both years with 33% and 40%.

In Norway Bjørnsson (1940) found type 1 in 22% and type 2 in 1% of cases. The same year Vammen (1940) in Denmark found type 1 in 21% and type 2 in 2% of cases.

TABLE V

**Typing within the pneumococcal groups of 384 strains found in blood, spinal fluid and pleural exudate during 1962–1969**

| Pn | Blood | Spinal fluid | Pleural exudate | Total | Pn | Blood | Spinal fluid | Pleural exudate | Total |
|----|-------|--------------|-----------------|-------|----|-------|--------------|-----------------|-------|
| 6A | 10 | 12 | 1 | 23 | 19F | 6 | 13 | 5 | 24 |
| 6B | 11 | 8 | 2 | 21 | 19A | 7 | 6 | 2 | 15 |
|    |    |   |   |    | 19B | — | — | — | — |
| 7F | 22 | 20 | 3 | 45 | 19C | — | — | — | — |
| 7A | — | — | — | — |    |    |   |   |    |
| 7B | — | — | — | — | 22F | 7 | 4 | 4 | 15 |
| 7C | 2 | — | 1 | 3 | 22A | 2 | — | 1 | 3 |
| 9A | 2 | 1 | — | 3 | 23F | 18 | 20 | 6 | 44 |
| 9L | — | 1 | 1 | 2 | 23A | 6 | 5 | 2 | 13 |
| 9N | 13 | 8 | 2 | 23 | 23B | — | 1 | — | 1 |
| 9V | 4 | — | 1 | 5 |    |    |   |   |    |
|    |    |   |   |    | 24F | 1 | — | — | 1 |
| 10F | 2 | 1 | 1 | 4 | 24A | — | — | — | — |
| 10A | 2 | 3 | 3 | 8 | 24B | 1 | — | — | 1 |
| 11F | 1 | 1 | — | 2 | 28F | 1 | — | 1 | 2 |
| 11A | 3 | 3 | 1 | 7 | 28A | 2 | 2 | 1 | 5 |
| 11B | 2 | 3 | — | 5 |    |    |   |   |    |
| 11C | — | — | — | — | 32F | — | — | — | — |
|    |    |   |   |    | 32A | — | 1 | — | 1 |
| 12F | 8 | 2 | — | 10 |    |    |   |   |    |
| 12A | 12 | 3 | — | 15 | 33F | 1 | 4 | — | 5 |
|    |    |   |   |    | 33A | 1 | — | — | 1 |
| 15F | — | 1 | — | 1 | 33B | — | 1 | — | 1 |
| 15A | 2 | 3 | 1 | 6 | 33C | 1 | — | — | 1 |
| 15B | 2 | 1 | 1 | 4 |    |    |   |   |    |
| 15C | 1 | 3 | 1 | 5 | 35F | 1 | — | — | 1 |
|    |    |   |   |    | 35A | — | 1 | — | 1 |
| 17F | 3 | 1 | 2 | 6 | 35B | — | 1 | 2 | 3 |
| 17A | — | — | — | — | 35C | — | 1 | 1 | 2 |
| 18F | 2 | 3 | — | 5 | 41F | — | — | — | — |
| 18A | 1 | 1 | — | 2 | 41A | — | 1 | — | 1 |
| 18B | 1 | 2 | — | 3 |    |    |   |   |    |
| 18C | 9 | 23 | 3 | 35 | Total | 170 | 164 | 49 | 384 |

In 1949, Lund reported on the frequency of pneumococcal types in Denmark during the years from 1939 to 1947. Similar investigations for 1955–1970 showed that type 1 had decreased from 13·1% to 1% and type 2 from 6% to 1·5%.

TABLE VI

**Distribution of pneumococcal types 1 and 2 in pneumonia at different times and in different areas.**

| Area | Year | Percentage | |
|------|------|--------|--------|
| | | Type 1 | Type 2 |
| USA | 1913 | 47 | 18 |
| USA | 1914 | 30 | 39 |
| South Africa | 1913–17 | 22 | 16 |
| Denmark | 1923 | 33 | 27 |
| Denmark | 1937 | 40 | 1·7 |
| South Africa | 1938 | 22 | 15 |
| Norway | 1940 | 22 | 1 |
| Denmark | 1940 | 21 | 2 |
| Denmark | 1939–47 | 13·1 | 6 |
| Denmark | 1955–70 | 1·0 | 1·5 |

## VIII. PNEUMOCOCCAL INFECTIONS IN ANIMALS

Pneumococcal mastitis is seen quite frequently in adult cattle. In calves, pneumococci often cause septicaemia, while monkeys — like man — seem to be able to limit the infection, which is usually restricted to the respiratory tract.

Guinea pigs have a particularly high susceptibility to infection with pneumococcal type 19F. In sheep and goats, cases of septicaemia and pneumonia are seen. Pigs seem to be fairly resistant to pneumococcal infections (Rømer, 1962).

The pneumococci found in animals and in man are identical and infection may be carried from animal to man and *vice versa*.

The pneumococci have constant virulence for mice as long as the capsule formation is unchanged. Pneumococcal cultures have been transferred in serum broth 300 times at 37 and 41 °C and showed no reduction in virulence as long as the capsule reaction was optimal. Thus, the virulence depends on the capsule and is constant for the same type. All S (smooth) pneumococci are virulent for mice, but to greatly varying degrees, while R (rough) pneumococci are avirulent (Mørch, 1943).

## IX. VACCINATION WITH TYPE-SPECIFIC POLYSACCHARIDES

It is not surprising that Wright *et al.* (1914) were unable to demonstrate with certainty that vaccination with heat-killed pneumococci could prevent pneumococcal pneumonia among South African miners, since their studies were carried out without knowledge about the variety of pneumococcal types with antigenically different polysaccharide capsules. Immunogenicity of such polysaccharides, however, was proved in 1930 by Francis and Tillett, and in 1945 MacLeod *et al.* delivered conclusive evidence that a pneumococcal vaccine consisting of four capsular polysaccharides induced type specific protection against epidemic pneumonia among recruits.

The introduction of antibiotics led to a diminishing interest in further work with pneumococcal vaccines but, unfortunately, it was found that pneumococci continued to cause life-threatening infections. In the U.S.A. more than 60,000 deaths and between 200,000 and 1 million cases of pneumonia a year are thought to be due to pneumococci (Austrian, 1974). In Denmark, the mortality from pneumococcal meningitis has remained virtually unchanged, at a level of 20%, during the last 30 years. Mortality is considerably higher in developing countries (Baird *et al.*, 1976) and lobar pneumonia the leading cause of hospitalization. Permanent hearing loss even in the Western Hemisphere is most often due to recurrent pneumococcal otitis media. The risk of acquiring—often fulminant—pneumococcal infection is greatly enhanced in splenectomized individuals (Gopal and Bisno, 1977; Likhite, 1977). Small wonder that considerable efforts have been devoted during the last decade to producing a new pneumococcal vaccine, the interest in which has been further aroused after the recent detection in South Africa of strains which are resistant to Penicillin (Appelbaum *et al.*, 1977).

By the end of 1977 licence was given in the U.S.A. to a vaccine composed of 14 highly purified pneumococcal polysaccharides produced from types 1, 2, 3, 4, 6A, 7F, 8, 9N, 12F, 14, 18C, 19F, 23F and 25. One dose consists of 50 $\mu$g of each polysaccharide. A number of studies have shown the vaccine to be safe, antigenic and 80%–90% effective in providing protection against type specific pneumonia (Austrian *et al.*, 1976; Smit *et al.*, 1977). Children under the age of 2 years, however, react unsatisfactorily (Austrian, 1976). Theoretically, to achieve maximal protection, all 83 polysaccharides should be used, but as certain types occur only infrequently, the 14 types included together cover approximately 85% of all isolates from cases of bacteraemia and meningitis in Denmark (Henrichsen, unpublished). Duration of the protection is as yet unknown, but the increased level of type specific antibodies appears to persist for at least three years and may be even longer. Re-vaccination does not lead to a booster-effect and apparently should not be performed at less than three-year intervals (Hilleman *et al.*, 1978). Studies are under way that are designed to shed light on the still unanswered question of whether or not pneumococcal vaccines will offer protection against recurrence of otitis media in children.

At present it is of major importance that side-effects are minimal and that the antibody response in splenectomized patients is comparable to that of otherwise healthy individuals (Ammann *et al.*, 1977; Sullivan *et al.*, 1978).

Vaccination of other populations at risk of acquiring infection, e.g. people above 50 years of age and those with chronic illnesses, should also receive consideration. Eventually, pneumococcal vaccines, together with other polysaccharide vaccines, will probably come into use for the prevention of bacterial meningitis.

## REFERENCES

Ammann, A. J., Addiego, J., Wara, D. W., Lubin, B., Smith, W. B., and Mentzer, W. C. (1977). *New Engl. J. Med.*, **297**, 897–900.

Anhalt, J. P., and Yu, P. K. W. (1975). *J. clin. Microbiol.* **2**, 510–515.

Appelbaum, P. C., Bhamjee, A., Schragg, J. N., Hallett, A. F., Bowen, A. J., and Cooper, R. C. (1977). *Lancet*, **2**, 995–997.

Austrian, R. (1973). *In* "New Approaches for Inducing Natural Immunity to Pyogenic Organisms" (J. B. Robbins, R. E. Horton and R. M. Krause, Eds), pp. 39–44. DHEW Publication No. (NIH) 74–553.

Austrian, R. (1974). *Prev. Med.*, **3**, 443–445.

Austrian, R. (1975). *J. infect. Dis.*, **131**, 474–484.

Austrian, R. (1976). *Mt. Sinai J. Med.*, **43**, 669–709.

Austrian, R., Douglas, R. M., Schiffman, G., Coetzee, A. M., Koornhof, H. J., Hayden-Smith, S., and Reid, R. D. W. (1976). *Trans. Ass. Am. Phys.* **89**, 184–194.

Baird, D. R., Whittle, H. C., and Greenwood, B. M. (1976). *Lancet*, **2**, 1344–1346.

Bjørnsson, J. (1940). *Nord. Med.*, **8**, 2082–2086.

Buchanan, R. E., and Gibbons, N. E. (Eds) (1974). "Bergey's Manual of Determinative Bacteriology", 8th edn. The Williams and Wilkins Co., Baltimore.

Bullowa, J. G. M. (1937). *J. Am. med. Ass.*, **109**, 2061–2069.

Christensen, S. (1923). Thesis. Arnold Busck, Copenhagen.

Colding, H., and Lind, I. (1977). *J. clin. Microbiol.*, **5**, 405–409.

Coonrod, J. D., and Rytel, M. W. (1972). *Lancet*, **1**, 1154–1156.

Eddy, B. E. (1944). *Publ. Hlth Rep. (Wash.)*, **59**, 449–452.

El-Refaie, M., and Dulake, C. (1975). *J. clin. Path.*, **28**, 801–806.

Engbaek, H. C. (1949). Thesis. Nyt Nordisk Forlag, Copenhagen.

Francis, Jr., T., and Tillett, W. S. (1930). *J. exp. Med.*, **52**, 573–585.

Gopal, V., and Bisno, A. L. (1977). *Arch. intern. Med.*, **137**, 1526–1530.

Griffith, F. (1928). *J. Hyg. (Lond.)*, **27**, 113–159.

Gundel, M. (1933). *Z. Hyg. Infekt.-Kr.*, **115**, 495–518.

Hayes, W. (1968). "The Genetics of Bacteria and Their Viruses", 2nd edn. Blackwell Scientific Publications, Oxford and Edinburgh.

Heidelberger, M., Jann, K., Jann, B., Ørskov, F., Ørskov, I., and Westphal, O. (1968). *J. Bact.*, **95**, 2415–2417.

Hilleman, M. R., McLean, A. A., Vella, P. P., Weibel, R. E., and Woodhour, A. F. (1978). *Bull. Wld Hlth Org.*, **56**, 371–375.

Kauffmann, F., and Langvad-Nielsen, A. (1942). *Acta path. microbiol. scand.*, **19**, 108–111.

Kauffmann, F., and Lund, E. (1954). *Int. Bull. bact. Nomencl.*, **4**, 125–128.

Kauffmann, F., Mørch, E., and Schmith, K. (1940). *J. Immunol.*, **39**, 397–426.

Kauffmann, F., Lund, E., and Eddy, B. (1960). *Int. Bull. bact. Nomencl.*, **10**, 31–40.

Likhite, V. V. (1976). *J. Am. med. Ass.*, **236**, 1376–1377.

Lister, F. S. (1913). Publ. Res. Div. South African Inst. Med. Res. No. II.

Lund, E. (1949). *Acta path. microbiol. scand.*, **26**, 83–92.

Lund, E. (1950). *Acta path. microbiol. scand.*, **27**, 110–118.

Lund, E. (1959). *Acta path. microbiol. scand.*, **47**, 308–315.

Lund, E. (1962). *Acta path. microbiol. scand.*, **56**, 87–88.

Lund, E. (1970a). *Int. J. system. Bact.*, **20**, 321–323.

Lund, E. (1970b). *Acta. path. microbiol. scand.*, **78**, 333–336.

Lund, E., and Munksgaard, A. (1967). *Acta path. microbiol. scand.*, **70**, 305–310.

Lund, E., Munksgaard, A., and Stewart, S. M. (1972). *Acta path. microbiol. scand.*, **B 80**, 497–500.

Lund, E., and Rasmussen, P. (1966). *Acta path. microbiol. scand.*, **68**, 458–460.

MacLeod, C. M., Hodges, R. G., Heidelberger, M., and Bernhard, W. G. (1945). *J. exp. Med.*, **82**, 445–465.

McGowan, Jr., J. E., Barnes, M. W., and Finland, M. (1975). *J. infect. Dis.*, **132**, 316–335.

Mørch, E. (1943). "Serological Studies on the Pneumococci", pp. 159–163. Humphrey Milford, Oxford University Press, London.

Mørch, E. (1944). *Acta path. microbiol. scand.*, **21**, 142–148.

Mørch-Lund, E. (1949). *Acta path. microbiol. scand.*, **26**, 709–714.

Mørch, E., and Knipschildt, H. E. (1944). *Acta path. microbiol. scand.*, **21**, 102–109.

Neufeld, F. (1902). *Z. Hyg. Infekt.-Kr.*, **40**, 54–72.

Nissen, N. I. (1937). *Ugeskr. Laeg.*, **99**, 801–805.

Ordman, D. (1938). *Publ. Res. Div. South African Inst. Med. Res.* No. XLII.

Pope, C. G., and Smith, M. L. (1932). *J. Path. Bact.*, **35**, 573–589.

Roesgaard, C. R. (1945). Thesis, University of Copenhagen.

Rømer, O (1962). Thesis. Carl F. Mortensen, Ltd., Copenhagen.

Schiffman, G., Summerville, J. E., Castagna, R., Douglas, R., Bonner, M. J., and Austrian, R. (1974). *Fedn. Proc. Fedn. Am. Socs. exp. Biol.*, **33**, 758.

Smit, P., Oberholzer, D., Hayden-Smith, S., Koornhof, H. H., and Hilleman, M.R. (1977). *J. Am. med. Ass.*, **238**, 2613–2616.

Sullivan, J. L., Ochs, H. D., Schiffman, G., Hammerschlag, M. R., Miser, J., Vichinsky, E., and Wedgwood, R. J. (1978). *Lancet*, **1**, 178–181.

Vammen, B. (1940). Thesis. Munksgaard, Copenhagen.

Wright, A. E., Morgan, P. W., Cantab, M. B., Colebrook, L., and Dodgson, R. W. (1914). *Lancet*, **1**, 1–10 and 87–95.

# Subject Index

## A

*Actinomyces viscosus*, 227
Agglutination reaction, 6
Antibodies,
  fluorescent, 10
  IgG, 11
  IgM, 11
Antigens,
  carbohydrate, 2
  demonstration of, 3
  genetic determination of, 3
  protein, 2

## B

*Bacillus cereus*, 227
*B. megaterium*, 227
Bacteria, serotyping of, 1–12
Bacteriolysis, 7
*Bacterium enterocoliticum*, 37
*Brucella*, 41

## C

Capsular polysaccharides, 4
Capsular swelling reaction, 5
Cholera,
  bacteriophage studies, techniques
    used in, 103
  bacteriophages,
    host range of, 41
    neutralisation of, 95
  El Tor, spread of, 79, 80
  group IV phage sensitivity, 61
  in India, 76
  inhibition of phages of, by specific
    antisera, 90
  phage type map, 77
  typing phages of,
    morphology of, 81
    plaque morphology of, 81
  vibrio biotypes, characteristics of, 55
*Clostridium botulinum*, toxins of, 10
*Cl. perfringens*, 232, 234
*Cl. septicum*, 234
Complement fixation reactions, 7

Complement, source of, 8
*Corynebacterium*, 227
*C. diphtheriae*, 98, 168, 234

## E

El Tor, 52 *et seq.*
*Enterobacter aerogenes*, 120
*E. cloacae*, 120
*Enterobacteriaceae*, 16, 17, 21, 33, 64,
  168, 169
Enterocins, 216, 231
Enterococci, 199 *et seq.*, 231, *et seq.*
  antigenic structure of, 203
  bacteriocin typing of, 215
  bacteriolysins of, 232
  characteristics of, 202
  classification of, 200
  definition of, 200
  differentiation of, 205
  D-specific antigen in, 203
  enterocin typing methods for, 216
  growth requirements of, 202
  identification of, 199–217
  identification methods for, 204
  peptidoglycan, 204
  phage-typing of, 214
  physiological characteristics of, 202
  selective media for, 204
  type-specific antigens in, 203
  typing of, 199–217
Enterococcins, 231
Enterococcus,
  typing of,
    principles of, 212
    serological, 213
*Escherichia coli*, 17, 38, 40, 41, 47, 119,
  123, 124, 169, 231, 246

## F

Fluorescent antibodies, 10
*Francisella tularensis*, 41
Freund's adjuvant, 11

## G

Gel diffusion reaction, 5